Synthesis Lectures on Communications

This series of short books cover a wide array of topics, current issues, and advances in key areas of wireless, optical, and wired communications. The series also focuses on fundamentals and tutorial surveys to enhance an understanding of communication theory and applications for engineers.

Xin Lin

Visible Light
Communications

 Springer

Xin Lin
Tokyo, Japan

ISSN 1932-1244   ISSN 1932-1708   (electronic)
Synthesis Lectures on Communications
ISBN 978-3-031-64474-0   ISBN 978-3-031-64475-7   (eBook)
https://doi.org/10.1007/978-3-031-64475-7

This Springer imprint is published by the registered company Springer Nature Switzerland AG
The registered company address is: Gewerbestrasse 11, 6330 Cham, Switzerland

If disposing of this product, please recycle the paper.

To my family

Preface

Visible light is an electromagnetic wave with wavelengths in a region of the spectrum from about 380 to 780 nm. The developments of visible lightwave technology, such as lightwave propagation, detection, and switching have found ever-increasing applications in optical communications, signal processing, computing, and sensing.

Visible light communication (VLC) as one of lightwave technology is a novel wireless communication technology developed along with LED lighting. One of its most important features is the ability to use the ubiquitous lighting to transmit information around the users, and this makes VLC one of the best techniques to realize ubiquitous information services, such as indoor short-range communications, Li-Fi (Light Fidelity) systems, IoT (Internet of Things) systems, underwater optical wireless communication systems and so on. Thus, it can also be referred to as illumination light communication (ILC), it is a modern and interesting technique related with the daily life of people.

The purpose of this book is to help readers understand the fundamentals that are related with this emerging technology, and its best application areas. Subject matter of this book has been chosen with two general aims in mind. The first is to give readers enough of the basic principles behind practical optical components and systems so that they can do effective research and laboratory works. The second aim is for readers to know the essence of VLC so they can find and build new applications.

This book, apart from emphasizing the basics and applications of VLC, also incorporates many of the recent developments, such as visible-light wireless LAN, underwater optical wireless sensor network, optical metamaterials, Optical multiple-access techniques, new VLC standards, and techniques of combining VLC with art. The book includes author's worked examples of the research and developments and a wealth of references. Depending on the requirement, topics in the book can as a research or development material for researchers and engineers in optical engineering or optical communication fields. And it also can provide an important reference material for either an introductory or a more advanced course to science students.

Tokyo, Japan   Xin Lin

Contents

Acronyms

ACO-OFDM	Asymmetrically Clipped Optical OFDM
ADC	Analog to Digital Conversion
AEL	Allowable Exposure Limit
AM	Amplitude Modulation
APD	Avalanche Photodiode
ARIB	Association of Radio Industries and Businesses
ASK	Amplitude Shift Keying
AUV	Autonomous Underwater Vehicle
B	Blue
BCSK	Binary CSK
BER	Bit Error Rate
BIM	Baseband Intensity Modulation
C	Cyan
CC	Convolutional Codes
CCD	Charge-Coupled Device
CD	Coherent Detection
CDM	Code Division Multiplexing
CIE	Commission Internationale de l'Éclairage (International Commission on Illumination)
CMOS	Complementary Metal Oxide Semiconductor
CMY	Cyan, Magenta, and Yellow
CRC	Cyclic Redundancy Checks
CRI	Color Rendering Index
CRT	Cathode Ray Tube
CSK	Color-Shift Keying
CSML	Color-Separation Metalens
CSS	Carrier Signal Source
D/A	Digital-to-Analog

DBM	Duobinary Modulation
DBR	Distributed Bragg Reflector
DC	Direct Current
DCO-OFDM	Direct Current Biased Optical OFDM
DD	Direct Detection
DEMUX	Demultiplexer
DLL	Data Link Layer
DMT	Discrete Multitone
DSOC	Deep Space Optical Communication
DSP	Digital Signal Processing
E/O	Electro-Optic Conversion
EEL	Edge-Emitting Laser
ELF	Extremely Low Frequency
EMC	Electromagnetic Compatibility
EPE	External Photoelectric Effect
FDM	Frequency Division Multiplexing
FEC	Forward Error Correction
FM	Frequency Modulation
FOC	Fiber Optic Communication
FOV	Field of View
FSK	Frequency Shift Keying
FSO	Free-Space Optics
FWHM	Full Width at Half Maximum
G	Green
GPS	Global Positioning System
HD	Heterodyne Detection
IC	Integrated Circuit
ID	Identification
IEC	International Electrotechnical Commission
IEEE	Institute of Electrical and Electronic Engineers
IFFT	Inverse Fast Fourier Transform
ILC	Illumination Light Communication
IM	Intensity Modulation
IM/DD	Intensity Modulation with Direct Detection
IoT	Internet of Things
IPE	Internal Photoelectric Effect
IR/Ir	Infrared
IrDA	Infrared Data Association
IS	Image Sensor
ISC	Image Sensor Communication
ISS	Information Signal Source

ITU	International Telecommunication Union
ITU-T	ITU Telecommunication Standardization Sector
JEITA	Japan Electronics and Information Technology Industries Association
JIS	Japanese Industrial Standards
LAN	Local Area Network
LBS	Location-Based Service
LD	Laser Diode
LED	Light-Emitting Diode
Li-Fi	Light Fidelity
LOS	Line of Sight
LTE	Long Term Evolution
M	Magenta
MAC	Medium Access Control
MAT	Multiple-Access Technique
MCSK	Multilevel CSK
MIMO	Multiple-Input-Multiple-Output
MLM	Multilevel Modulation
MUX	Multiplexer
NIR	Near Infrared
NRZ	Non Return to Zero
O/E	Optic-Electro Conversion
OCC	Optical Camera Communication
OCDM	Optical Code Division Multiplexing
OCR	Optical Clock Rate
ODBM	Optical Duobinary Modulation
OEIC	Optoelectronic Integrated Circuit
OFDM	Orthogonal Frequency Division Multiplexing
OHD	Optical Heterodyne Detection
OMLM	Optical Multilevel Modulation
OMTA	Optical Multiple-Access Technique
OOK	On-Off Keying
OOK-NRZ	OOK with NRZ
OTDM	Optical Time Division Multiplexing
OWAP	Optical Wireless Access Point
OWC	Optical Wireless Communication
OWT	Optical Wireless Transmission
OWTC	Optical Wireless Transceiver
PAM	Pulse-Amplitude Modulation
PAPR	Peak-to-Average Power Ratio
PCC	Photon-Counting Camera
PCD	Photon-Counting Detector

PD	Photodiode
PDM	Polarization Division Multiplexing
PHY	Physical
p-i-n	Positive-Intrinsic-Negative
PIN-PD	Positive-Intrinsic-Negative Photodiode
PLC	Power Line Communication
PM	Phase Modulation
PM-QAM	Polarization-Multiplexed Quadrature Amplitude Modulation
PMT	Photomultiplier Tube
p-n	Positive-Negative
PPM	Pulse-Position Modulation
PSK	Phase Shift Keying
PWM	Pulse-Width Modulation
QAM	Quadrature Amplitude Modulation
R	Red
RGB	Red, Green, and Blue
RLL	Run Length Limited
ROV	Remotely Operated Vehicle
RS	Reed-Solomon
SC	Subcarrier
SCF	Subcarrier Frequency
SCM	Subcarrier Modulation
SDM	Space Division Multiplexing
SNR	Signal-to-Noise Ratio
SOWC	Space Optical Wireless Communication
TDM	Time Division Multiplexing
TV	Television
UOWC	Underwater Optical Wireless Communication
UOWSN	Underwater Optical Wireless Sensor Network
UV	Ultraviolet
VCSEL	Vertical Cavity Surface Emitting Laser
VL	Visible Light
VLC	Visible Light Communication
VLCA	Visible Light Communication Association
VLCC	Visible Light Communication Consortium
VLD	Visible Light Laser Diode, Visible Laser Diode
VPPM	Variable Pulse-Position Modulation
WAP	Wireless Access Point
WCDM	Wideband Code Division Multiplexing

WDM	Wavelength Division Multiplexing
Wi-Fi	Wireless Fidelity
WLAN	Wireless Local Area Network
Y	Yellow

Overview of Visible Light Communication

1

This chapter describes the physical nature and characteristics of visible light (VL) waves as the carrier for visible light communication (VLC), as well as why, in today's context where radio communication technologies are becoming increasingly sophisticated and widely applied, there is still a need to research and apply visible light communication technology.

Visible light communication (VLC) is a novel and attractive subject not only involving old basic optical theory including generation, propagation, and detection of light, but also involving new disciplines such as semiconductor optoelectronics and lightwave-modulation technology. Two major developments have been achieved in the last twenty years, are responsible for VLC increasing importance in modern technology: the introduction of the LED (light emitting diode) lighting, and high-speed semiconductor photodetector. As a result of these developments, new applications have emerged: terrestrial ubiquitous information services based on the LED lighting including indoor/outdoor illumination-light communication (ILC), Li-Fi, and LED-based IoT as well as non-terrestrial wireless communication including underwater and space optical wireless communication (UOWC and SOWC). Figure 1.1 shows the main enabling technologies and application fields related to VLC.

Especially, the rapid advance and diffusion of portable terminals in human life is accelerating the introduction of various information service methods by using wireless communication techniques. How to realize high-speed, long-distance, and *last 3-m connectivity* for ubiquitous data transmission are three different fields to construct a wireless communication system, and satisfying these requirements has motivated recent interest in VLC. This book aims at describing the basic principle, enabling techniques, and current applications for VLC, its structure proceeds as follows.

© The Author(s), under exclusive license to Springer Nature Switzerland AG 2025   1
X. Lin, *Visible Light Communications*, Synthesis Lectures on Communications,
https://doi.org/10.1007/978-3-031-64475-7_1

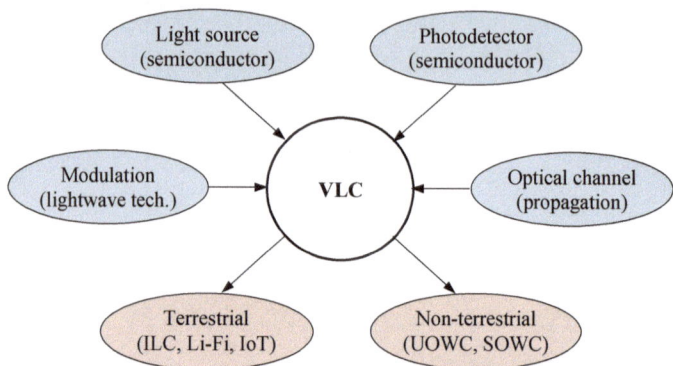

Fig. 1.1   Main enabling technologies and application fields related to VLC

This chapter is an overview and gives a comprehensive description of VLC. Chapters 2 and 3 focus on the key components of the VLC system. Basic principles and various characteristics of the light source and detector, which are the main components of VLC, are explained in detail. Chapters 4 to 6 discuss the contents relevant to system including methods of VLC system design and of analysis of spatial channel, various modulation schemes for VLC, as well as optical multiplexing techniques. Chapter 7 describes VLC standardization and several current major standards for VLC. Finally, Chap. 8 is devoted to current applications of VLC.

I hope readers will enjoy reading this book. By which I would like to express my thanks to my colleagues, without their hard work VLC would not make today's progress.

1.1   What Is Visible Light

Visible light (VL) is a segment of electromagnetic radiation that most human eyes can view. Electromagnetic radiation is transmitted in waves and particles at different wavelengths and frequencies.

The wavelength range of electromagnetic radiation is known as the electromagnetic spectrum. The electromagnetic spectrum is typically divided into seven regions in order of increasing wavelength and decreasing frequency, and the common designations are γ-rays, X-rays, UV (ultraviolet), VL (visible light), IR (infrared), microwaves, and radio waves. The range of optical wavelengths contains three bands that in UV, VL, and IR. VL falls between UV and IR has wavelength range from about 380–780 nm, and the corresponding frequencies is about 790–405 THz. Figure 1.2 illustrates the frequencies and wavelengths of electromagnetic radiation and, the position of VL band in broad electromagnetic spectrum. We see VL waves as the colors of the rainbow. Each color has a

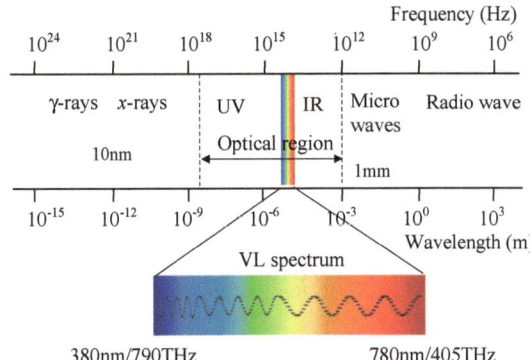

Fig. 1.2 Frequencies and wavelengths of electromagnetic wave

different wavelength. Violet light ray has the shortest wavelength, at around 380 nm, and red has the longest wavelength, at around 780 nm.

1.1.1 Nature of VL

British scientist Isaac Newton first uses the word *spectrum* (Latin for *appearance* or *apparition*) to describe the experiment phenomenon in his book Optics in the seventeenth century. He in 1666 passed sunlight (white light) through a narrow slit and then a glass prism to project the optical spectrum onto a wall, and he discovered that the glass prism could disassemble and reassemble white light. his experiment showed that when the full spectrum of VL through a prism, the wavelengths are separated into the rainbow colors because each wavelength has a different color, and each monochromatic light ray is bent at a slightly different angle depending on the wavelength of the color because their different refractivity, as shown in Fig. 1.3.

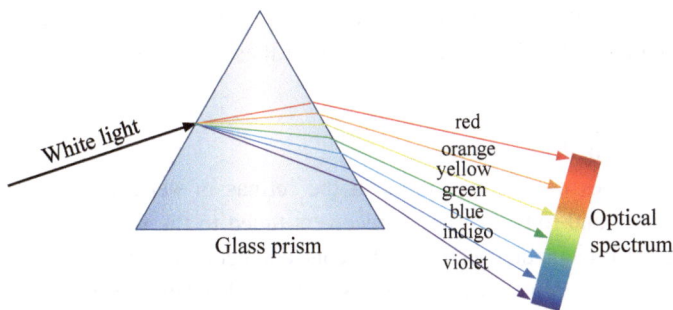

Fig. 1.3 Generation of VL spectrum by using a glass prism

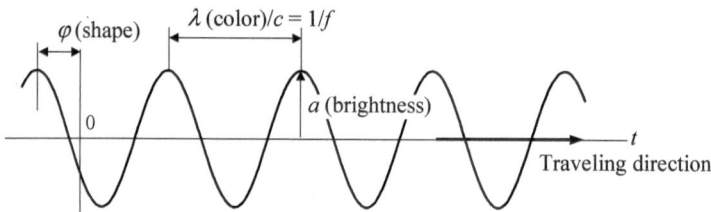

Fig. 1.4 Representations of a monochromatic wave at a fixed position

VL propagates in the form of transverse waves. A monochromatic wave at a fixed position can be represented by a wavefunction with harmonic time dependence

$$E(t) = a \cdot \cos[2\pi f t + \varphi], \tag{1.1}$$

as shown in Fig. 1.4, where a is amplitude, j is phase, and f (cycles/s or Hz) is frequency. In free space, light waves travel with a constant speed $c = 3.0 \times 10^8$ m/s. The light wavefunction is also therefore related to light wavelength l by

$$E(t) = a \cdot \cos\left[2\pi \frac{c}{\lambda} t + \varphi\right]. \tag{1.2}$$

For VL wavefunction, the light amplitude denotes an object's brightness, the wavelength indicates the color, and the phase corresponds to the shape. Hence, *visible light* can describe full picture of an object.

1.1.2   VL and Color

The most important characteristic of VL is color. Color is both an inherent property of light and an artifact of the human eyes. The color of light depends on its wavelength range from 380 nm at the violet end of the spectrum to 780 nm at the red end, as shown in Fig. 1.3, and the color human see is a result of these wavelengths are reflected back to the eyes.

Spectral Response of Human Eyes

There are two types of photoreceptors in the retinas of vertebrate eyes including the human eye, cones and rods, that act as receivers tuned to the wavelengths in this narrow VL band of electromagnetic spectrum. The cones detect color and the rods only let us see things in black, white, and grey. The cones only work when the light is bright enough, this is why things look grey and we cannot see colours at night when the light is dim.

On the other hand, the human visual system does not respond uniformly to all wavelengths in the VL spectrum. It is most responsive in the middle portion of the spectrum

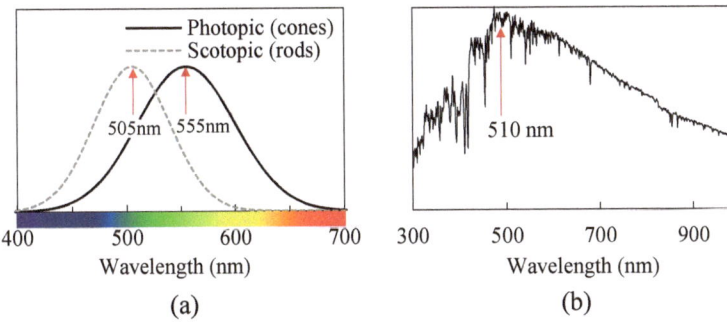

Fig. 1.5 Relative spectral response for **a** photopic and scotopic vision of the human visual system, and **b** solar light

and less responsive for red and blue wavelengths. Furthermore, the rods and cones have different spectral response curves, as shown in Fig. 1.5a.

Figure 1.5a shows response curves for photopic V(l) (solid line) and scotopic V'(l) (dashed line) vision, also called luminosity curves or luminous efficiency curves describes the average spectral sensitivity of human visual perception of brightness. The abscissa is wavelengths, and the ordinate represents relative spectral sensitivities. It is based on subjective judgements of which of a pair of different-colored lights is brighter, to describe relative sensitivity to light of different wavelengths.

Under photopic conditions of luminance >3 cd/m^2, the rods are saturated and only the cones are producing a visual signal. Under scotopic conditions of luminous <0.03 cd/m^2, the light levels are too low to activate the cones, but the rods still respond. Moreover there is also mesopic conditions that luminance range in 0.03–3 cd/m^2 refer to the in-between state where both rods and cones are active. The peak of the photopic curve occurs at a wavelength of 555 nm. The peak shifts to a wavelength of 505 nm for scotopic conditions. This shift in sensitivity towards the blue end of the spectrum in dim illumination is called the *Purkinje shift*.

As a comparison Fig. 1.5b gives the spectral response of the sunlight, it indicates an interesting result that the most sensitive wavelength range of the sunlight is also in the VL band, and the peak occurs at a wavelength of about 510 nm coincides with the peak of the human eye in Fig. 1.5a. In this sense, for humans who have evolved in the solar system, *light* is indeed a special important physical phenomenon closely related to *humans* and it is also one of the most precious treasures given to humans by nature. In fact, the sun is the dominant source for VL waves human eyes receive.

CIE 1931 Color Space

The CIE (Commission Internationale de l'Éclairage) defined photopic V(λ) as a standard function in 1924 and may be used to convert radiant energy into luminous (i.e., VL) energy. It also forms the central color matching function in the CIE 1931 color space

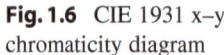

Fig. 1.6 CIE 1931 x–y
chromaticity diagram

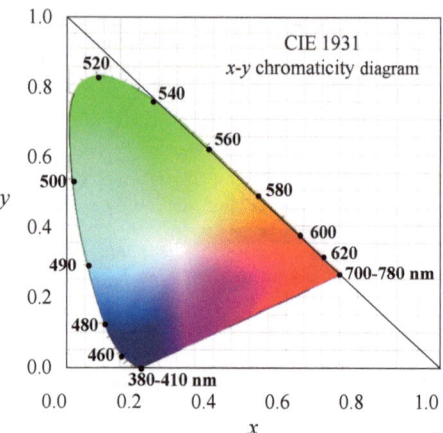

that first defined the quantitative links between distributions of wavelengths in the visible
spectrum, and physiologically perceived colors in human color vision. Figure 1.6 shows
the CIE 1931 chromaticity diagram.

The color values of light are represented by the plane coordinates of x, y. The light
wavelengths are written around the curve (i.e., spectral locus) of the chromaticity diagram. Each monochromatic is represented by these wavelengths. Using the chromaticity
diagram, the relations between the colored lights of red (R), green (G), blue (B) and the
coordinate values of x and y can be written by

$$\begin{cases} x = 0.6R - 0.28G - 0.32B \\ y = 0.2R - 0.52G + 0.31B \end{cases}. \tag{1.3}$$

Three Primary Colors of VL

The three primary colors of VL are RGB (red, green, and blue). Mixing these colors in
different proportions can make all the colors of the light are saw by human eyes. Because
the brightness will be up when any two of RGB are mixed together, it is called *additive
color mixing*, as shown in left of Fig. 1.7.

In the additive mixing space, cyan (C) is a mixture of green and blue, magenta (M) is
blend of red and blue, and yellow (Y) contains both red and green. Mixing lights of RGB
with equal proportions will generate white light. Black is a total absence of light. This
is how cathode-ray-tube (CRT) monitors and television (TV) screens work. For example,
red and green lights are used to make human brain perceive the image as yellow.

On the other hand, this is different when inks (or paints) are mixed. Each color of ink
is absorbing certain colors and reflecting others. Each time another color of ink is mixed
in, there are more colors absorbed and less are reflected. The three primary colors for
adding inks, such as for a printer, are cyan, magenta and yellow. The CMY is also called

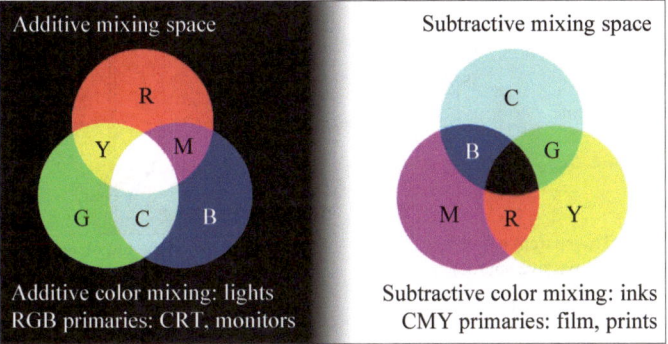

Fig. 1.7 Additive and subtractive color mixing spaces

richromatic. Since the brightness will be down when any two of C, M, and Y are mixed together, it is called *subtractive color mixing*, as shown in right of Fig. 1.7.

In the subtractive mixing space, R, G, and B corresponds to mix M and Y, C and Y, and C and M, respectively. If mixing equal amounts of C, M, and Y, all the lights will be absorbed and only see black color, because no light will be reflected back to human eyes.

Color of Objects

Strictly, objects don't have color, rather, they give off light that *appears* to be a color. objects appear different colors because they absorb some colored-lights wavelengths and reflect or transmit other one simultaneously. In fact, the colors we see are the light wavelengths that are reflected or transmitted. Therefore, relative to the active light source that emits light from itself such as sun, laser, and LED, an object can be considered as a passive light source which carries the active light source. The Sun with full-color light is the dominant active source for VL waves human eyes receive.

Figure 1.8 gives an example that how human eyes see red color when active source is a white light source with full colors. In Fig. 1.8a, observed object (i.e., passive source) is a transparent red glass. Light can pass through the transparent object (for translucent material object such as frosted glass or natural seawater with turbidity, partial light can pass through) and allows we to clearly see it on the other side. A red glass looks red because red light is the only light that is transmitted, and all the others have been absorbed. In Fig. 1.8b, observed object is an opaque red material which no light can pass through. It looks red because it only reflects red light and absorbs all the others. If only blue light is shone onto a red object, the object would appear black, because the blue would be absorbed and there would be no red light to be reflected or transmitted.

White objects appear white because they reflect all colors. Black objects absorb all colors, so no light is reflected.

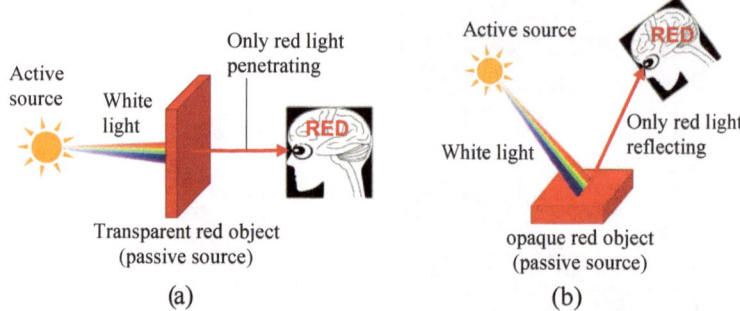

Fig. 1.8 Red light detecting for **a** transparent and **b** opaque object

1.2   What Is VLC

VLC employs VL for transmission both digital and analog signals. VLC systems usually work in the method of wireless communication, so it is a subset of optical wireless communication technologies.

1.2.1   History of VLC

VLC can be divided into three historical stages that are visual, natural light, and artificial light communications [1, 2].

Visual Communications
In ancient times, VL was used to convey visual messages for far away. About in the seventh century B.C., people used fire and smoke signals to convey situation oneself each other which has been written of in the historical document of ancient China *Shiji*. Figure 1.9a is an example for a set of smoke signals, using the number of the smoke to express different visual messages.

In order to convey messages for more longer distance, Claude Chappe invented the semaphore communication based on optical method in the late eighteenth century in France was used to communicate between the cities. The semaphore is a system of sending messages by holding the arms or two flags or poles in certain positions according to an alphabetic code, as shown in Fig. 1.9b. It uses different directions of the semaphore to construct a variety of information codes, and then used an optical telescope to read the semaphore signals.

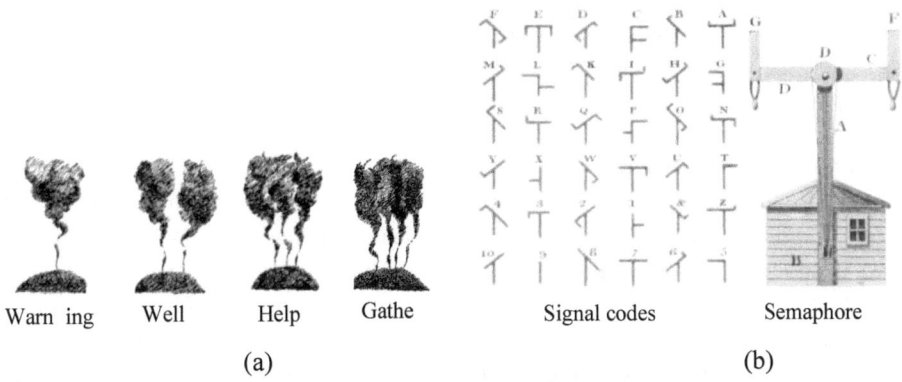

Warn ing   Well   Help   Gathe   Signal codes   Semaphore

(a)   (b)

Fig. 1.9 Visual communication methods using: **a** smoke signals and **b** semaphore

Natural Light Communications

The First Successful practical scheme for VLC can be attributed to the great inventor Alexander Graham Bell [3]. In 1880, A. G. Bell invented the photophone by using natural sunlight, a device in which speech could be transmitted on a beam of light as shown in Fig. 1.10.

In this scheme, a narrow beam of sunlight via a plane mirror and a lens was focused onto a flexible mirror. Original sound waves via a speaking tube striking the mirror caused it to vibrate with the intensity of the reflected light varying proportionally to the strength of the sound. A selenium (Se) detector whose resistance varied according to the intensity of incident light then detected the light [4]. In this way, Bell managed to send a voice signal over distance of about 200 m. The major drawback using natural light for communications is that the system's work depends on the weather.

Though Bell's photophone did not reach commercial fruition due to the lack of a dependable light source and of a low-loss transmission medium, it shows that the medium

Fig. 1.10 Schematic of Bell's photophone

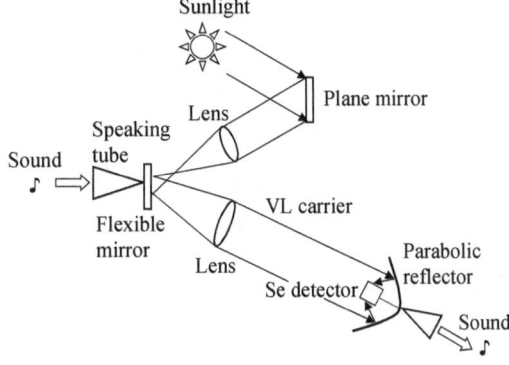

for information transmissions can use not only long-wavelength radio wave but also short-wavelength VL wave, and it forms an important cornerstone for VLC. In Bell's own words: "I regard the photophone as the greatest invention I have ever made, greater than the telephone" [1].

Artificial Light Communications

Fortunately, in the long-term social activities, human developed various artificial light sources in order to conquer the darkness and expand the living space oneself. Figure 1.11 shows historical development in lightings and their luminescence mechanism.

As first generation (1G) lightings the candles were among the earliest inventions of the ancient world, as shown by candlesticks from Egypt and Crete dating to at least 3000 BC. In the nineteenth century a French chemist, Michel-Eugène Chevreul, separated the fatty acid from the glycerine of fat to produce stearic acid, from which superior candles could be made. In use, heat from the flame liquefies the wax near the base of the wick. The liquid flows upward by capillary action, then is vaporized by the heat. The flame is the combustion of the wax vapour, as shown in Fig. 1.11a. The candlelight is one of candoluminescence. The Standard, or International, Candle is a measurement of light source intensity.

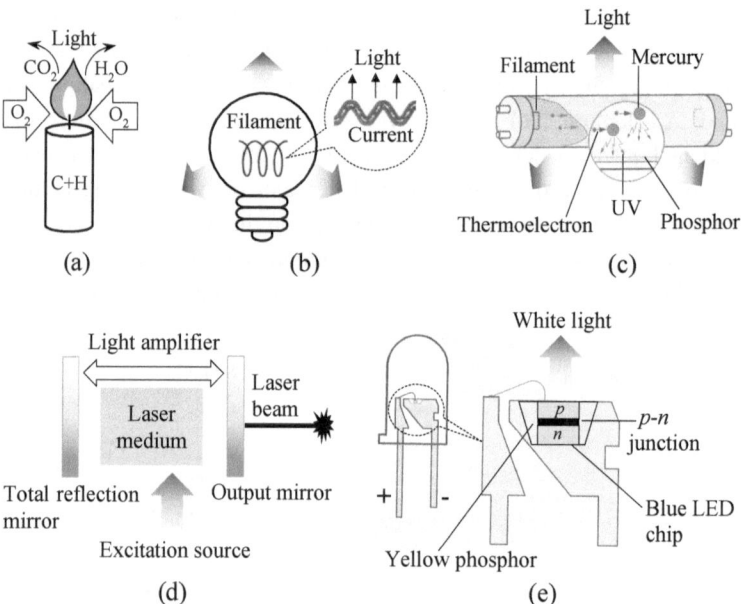

Fig. 1.11 Historical development in light sources: **a** candle, **b** incandescent lamp, **c** fluorescent lamp, **d** laser, and **e** white LED

Incandescent bulbs based on the method of electrothermal luminescence are 2G light-ings work by using electricity and a filament, as shown in Fig. 1.11b. Heated by electricity, the filament inside the light bulb exhibits resistance that results in high temperatures that cause the filament to glow and emit light. In 1802, Humphry Davy created the first incan-descent light by passing current through a platinum strip. It caused a glow and did not last long but marked the beginning of incandescent light development. In 1878 Amer-ican inventor, Thomas Edison began by tackling the problem of creating a long-lasting incandescent lamp, something that would be needed for indoor use. In 1921, incandescent lamps became the measurement standard of light intensity, and candles are no longer used for reference.

3G lighting is fluorescent lamp. The fluorescent light is a type of electric lamp by using the techniques of the electroluminescence that excites mercury vapor to create lumines-cence, as shown in Fig. 1.11c. At the turn of the twentieth century Peter Cooper Hewitt, an American electrical engineer, developed the basis for the modern fluorescent light. His low-pressure mercury arc lamp is the very first prototype of today's modern fluorescent lights. In an attempt to get more energy and less heat from an incandescent lamp, he invented "The Cooper Hewitt mercury-vapor lamp," which resulted in electrical currents passing through mercury gas sealed in a tube—the basis from which fluorescent lights operate.

Laser is a device that emits light through a process of optical amplification based on the stimulated emission of electromagnetic radiation, as shown in Fig. 1.11d. Excited photons enter the laser medium and then are reflected back and amplified by a mirror to form a laser beam. The first laser was built in 1960 by Theodore H. Maiman at Hughes Research Laboratories, based on theoretical work by Charles Hard Townes and Arthur Leonard Schawlow. The laser differs from other light sources in that it emits lights which is coherent, so has quite high emission-power density. This characteristic makes laser not suitable for lighting, but more suitable as the light source of a communication system. In fact, the popularity of optical fiber communications is due to the developments of laser technology.

Practical VLC had to wait until the arrival of the semiconductor diode source of the LED, as shown in Fig. 1.11e. In the 1990s, the success of blue LED chips made white LED lighting possible, at the same time, lighting has entered the 4G era. And VLC reached a new height when the LED lighting was applied successfully to indoor data transmission [5–8]. In 2011, IEEE (The Institute of Electrical and Electronic Engineers) developed the first standard for the use of VLC for wireless personal area networks—IEEE Std 802.15.7™-2011—covering topics such as network topologies, addressing, collision avoidance, acknowledgement, performance quality indication, dimming support, visibility support, colored status indication and color-stabilization [9]. It has driven VLC to make tremendous progress.

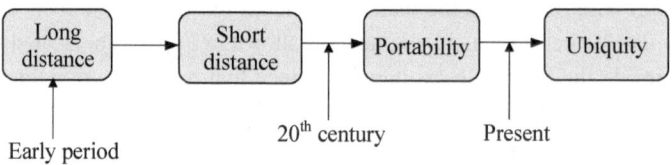

Fig. 1.12 Historical vicissitudes of wireless communications

1.2.2 Why VLC

Figure 1.12 shows historical vicissitudes of wireless communications. Initial wireless communication used to be an expensive technology. At the early twentieth century, the price of a wireless communication equipment was about equivalent to a battleship, so at that time, personal or short-distance communications almost impossible to use it, unless in case for long-distance or important military occasions.

Since the 1980s, developments of semiconductor and digital technologies have reduced the cost of wireless communication devices and make them to enter the commercial market gradually. As a result, wireless communication devices are used for personal or short distance have become possible.

In the twenty-first century, wireless communications have become an important technique for personal terminals and IoT, and it pursues portability and ubiquity. It expects that there is an information-service environment around each of us for everyday life.

On the other hand, in wireless communication systems using electromagnetic wave as a medium, the transmitter superimposes information on a carrier signal. At the receiver end, this information is retrieved from the carrier and processed. Theoretically, increasing the carrier frequency therefore increases the transmission bandwidth. VL band has frequency spectrum of 10^{14} Hz order (Fig. 1.2), which can thus support wideband modulation for optical communication.

The use of VL radiation as the medium for wireless communication offers several significant advantages over radio:

- VL emitters are available at low cost because the LED lighting can be used as a sending source.
- The VL spectral region offers a virtually unlimited bandwidth that is unregulated worldwide.
- VL radiation is absorbed by dark objects, diffusely reflected by light-colored objects, and directionally reflected by shiny surfaces. It can also penetrate through transparent and translucent objects such as glass, water and so on, but not through walls or other opaque barriers. These inherent characteristics of VL make VLC transmissions easy to secure against casual eavesdropping, and it prevents interference between links operating in different rooms.

Table 1.1 Comparison between radio and VL wireless links

Property of medium	Radio	VL	Implication for VL
Bandwidth regulated?	Yes	No	– Approval not required – Worldwide compatibility
Passes through walls?	Yes	No	– Short range – More easily secured
Passes through water?	Extremely low frequency only	Yes	– Simpler underwater wireless communication
Main noise?	Other users	Background light	– Difficult to operate outdoors
Input represent	Amplitude	Power	– High power requirement – Simpler intensity modulation

However, the VL medium is not without drawbacks compared to radio:

– Because VL cannot penetrate walls, communication from one room to another requires the installation of light access points that are interconnected via a wired backbone.
– In many indoor environments, there exists stronger ambient light noise, arising from sunlight, incandescent lighting, and fluorescent lighting, which induces noise in a VL receiver.
– At present, intensity modulation with direct detection (IM/DD) is most practical information transmission technique for VLC. IM/DD must employ relatively high transmit power for high signal-to-noise ratio (SNR) of a direct-detection receiver. However, when using LED lighting as a sending light source, too-strong power (brightness) will make people feel physiological discomfort. Often, the transmitter power level may be limited by concerns of power consumption and eye safety.

The properties of radio and VL wireless links are compared in Table 1.1 [10]. Radio and VL are complementary transmission media, and different applications favor the use of one medium or the other. Radio is favored for applications in which user mobility must be maximized or where transmission though walls or over long range is required and may be favored when transmitter power consumption must be minimized. For indoor ubiquitous, short-range underwater links, an international compatibility is required, however, and VLC is an attractive possibility.

References

1. S. Endo: *History of Light* (Tosho, Tokyo 1977)
2. A. Selvarajan, S. Kar, T. Srinivas: Overview of optical fiber communications. In: *Optical Fiber Communication*, ed. By G. Kelser (McGraw-Hill, New York 2002)
3. A.G. Bell: The photophone. Science, **os-1**(12), 130–134 (1880)
4. T. Aruga: *Spatial Transmission Optics* (Suiyosha, Tokyo 2000)
5. Nakagawa Laboratories: Illuminative light communication device, JP3827082 (2006) and US7583901 B2 (2009)
6. S. Haruyama: Visible light communications. IEICE Trans. **J86-A**(12), 1284–1291 (2003)
7. M. Nakagawa: Ubiquitous visible light communications, IEICE Trans. **J88-B**(2), 351–359 (2005)
8. X. Lin: Optical wireless ubiquitous information service using LED lighting, Mon. disp. **18**(10), 46–52 (2012)
9. IEEE *Standard 802.15.7-2011: Short-Range Wireless Optical Communication Using Visible Light* (IEEE, Piscataway 2011)
10. .M. Kahn, J.R. Barry: Wireless infrared communications, Proc. IEEE, **85**(2), 256–298 (1997)

The commercialization of visible light emitting diodes (LEDs) has rapidly led to the LED-ification of lighting, creating a favorable development environment and research value for VLCs primarily aimed at using illumination light for communications. In this chapter, two white LED light sources that can be used for both illumination and communication are introduced. Additionally, with the development of laser lighting technology in recent years, VL laser diodes (LDs) will also become a reasonable choice for VLC light sources. Therefore, this chapter also introduces VL LDs including vertical cavity surface emitting laser (VCSEL). Finally, a comparison is made between LEDs and LDs in terms of eye safety, output and modulation characteristics.

As described in Chap. 1, most current optical communication systems use light wave from artificial light source as the communication carrier to transmit data. The main light sources for VLC are LED and LD (laser diode).

Light sources for VLC should satisfy some desirable properties in terms of intensity, radiation pattern, emission wavelength, spectral characteristics and response time. Semiconductor diode sources of LED and LD types are a natural choice for VLC systems due to low power consumption, long life, compact size, rapid response (on/off time of devices), and regular spectral characteristic with single-wavelength emission.

An LED is basically an incoherent source. It emits radiation over a wide angle and contains a broad spectrum of wavelengths. On the other hand, an LD is highly coherent source, emits in a narrow range of angles, has a narrow spectrum and faster response time than LEDs. However, in terms of operating current requirement, cost and reliability, LEDs scores over LDs.

© The Author(s), under exclusive license to Springer Nature Switzerland AG 2025   15
X. Lin, *Visible Light Communications*, Synthesis Lectures on Communications,
https://doi.org/10.1007/978-3-031-64475-7_2

2.1   Light Emitting Diode (LED)

2.1.1   Basic Principles

The LED is a two-lead semiconductor light source. It is a positive–negative (p–n) junction diode under forward bias that emits light when activated, and consists of a chip of semiconducting material doped with impurities to create a p–n junction. When a suitable voltage is applied to the leads, electrons are able to recombine with electron holes within the device, releasing energy in the form of photons. This effect is called electroluminescence. The wavelength of the light emitted, i.e., color of the light, is determined by the energy band gap of the materials forming the p–n junction. The materials used for the LED have a direct band gap between Fermi and valence level with energies corresponding to near-infrared, visible, or near-ultraviolet light. The inner workings of an LED as shown in Fig. 2.1.

On the other hand, Electroluminescence is an optical phenomenon, and electrical phenomenon where a material emits light in response to an electric current passed through it. As the forward voltage increases, the intensity of the light increases and reaches a maximum. Figure 2.2 shows a simple LED electronic circuit. The circuit consists of an LED, a voltage supply and a resistor to regulate the current and voltage. The LED symbol is the standard symbol for a diode which has an anode and a cathode, with the addition of two small arrows denoting the emission of light.

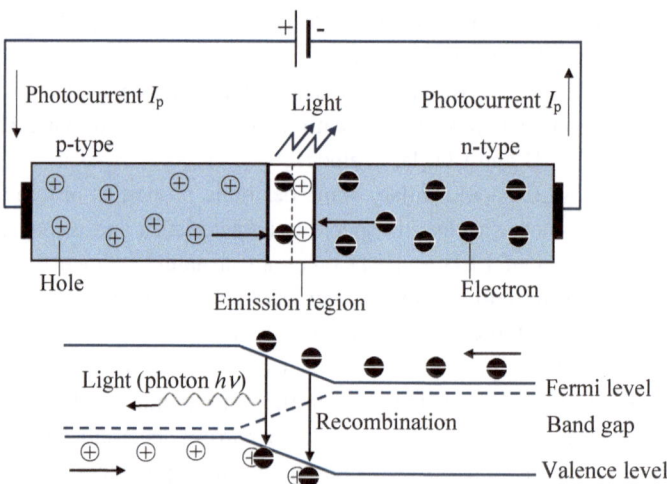

Fig. 2.1 Working principle of LED

Fig. 2.2 LED circuit

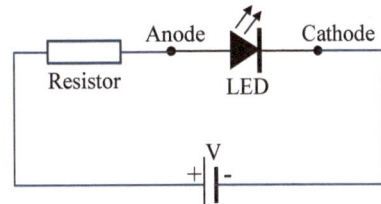

2.1.2 Development History

Figure 2.3 shows the development history of LED, it spans several decades began with infrared (IR) and red devices made with gallium arsenide (GaAs), their emission intensity is in the milliwatt order. Advances in materials science have enabled devices with ever-shorter wavelengths, emitting visible light in a variety of colors.

One of the critical milestones in LED development was the invention of the blue LED using indium gallium nitride (InGaN) in 1993. The breakthrough of Blue LED completed the primary colors needed for full-color displays and lighting applications.

White light LEDs by mixture of monochromatic blue LEDs and phosphor material have been able to achieve very high brightness, which over the watt order. This breakthrough opened up possibilities for LED lighting applications, offering energy efficiency, long lifespan, and durability. Now white LED has used for general illumination. Table 2.1 shows some available colors with wavelength range, voltage drop, and material [1].

LEDs offer numerous advantages, such as lower power consumption, longer lifespan, instant on/off, compact size, and environmental friendliness. The development of LED technology not only has revolutionized the lighting industry, providing energy-efficient alternatives to traditional incandescent and fluorescent lighting sources but also contribute to the advancement of emerging technologies. Their instant on/off and environmental friendliness make them best sources for VLC in cases of indoor or biological environment.

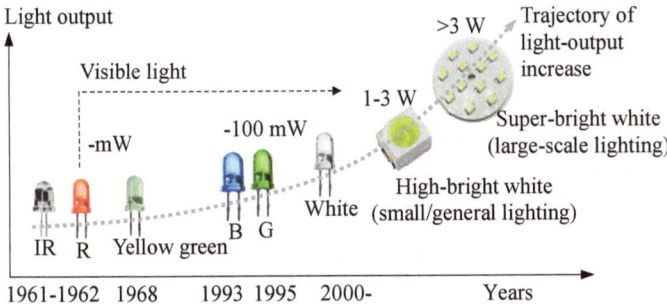

Fig. 2.3 The development history of LED

Table 2.1 Luminous colors corresponding to some semiconducting luminescent materials

Luminous colors	Wavelength range (nm)	Drive voltage (V)	Luminescent material
Infrared	>760	<1.63	GaAs, AlGaAs
Red (R)	610–760	1.63–2.03	AlGaAs, GaAsP, AlGaInP, GaP
Green (G)	500–570	1.9–4.0	GaP, AlGaInP, AlGaP, InGaN, GaN
Blue (B)	450–500	2.48–3.7	ZnSe, InGaN, SiC, Si
Ultraviolet (UV)	<400	3.0–4.1	InGaN, AlN, AlGaN, AlGaInN
White	Broad spectrum	2.8–4.2	– Color mixing by RGB LED – UV LED + RGB phosphor – Blue LED + yellow phosphor

2.1.3 Brightness Units

Brightness units used in LED specifications are candelas/millicandelas (cd/mcd), lumens/millilumens (lm/mlm), and watts/milliwatts (W/mW). Figure 2.4 shows their interrelationship in an LED.

The units cd/mcd represent luminous intensity at a specific forward current and the units are usually expressed as a value relative to the viewing angle. The larger the viewing angle, the more light flows given the same intensity, since light is not always evenly dispersed, and the output of light is determined by the location of the beholder. For example, 1 cd over 120° viewing angle is a lot brighter than 1 cd over 20° viewing angle.

The units lm/mlm are total luminous flux flowing from the LED; it is the power of light as perceived by the human eye with respect to the wavelength of light being emitted. If the luminous intensity (cd) and half viewing angle θ (degrees) are known, first converting θ from degrees to steradian (sr) by using sr $= 2\pi(1 - \cos\theta)$, and then luminous flux can

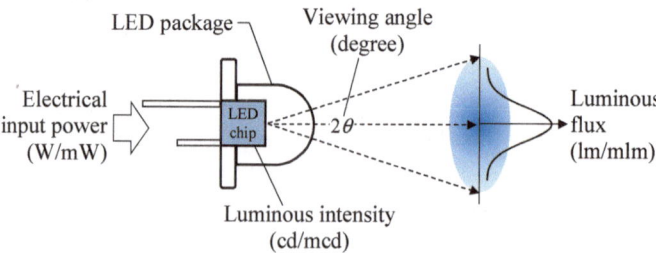

Fig. 2.4 The interrelationship of brightness units for LED

be given by lm = cd × sr or mlm = mcd × sr. lm/mlm units usually refer to the total light output of the device such as an LED lamp at the rated current.

The units W/mW represent the electrical power providing to the LED. The wattage of LED lamp or/and incandescent lamp is how much power consumption that particular lamp will draw, and is not necessarily equal to the power output of the lamp. That is why a smaller wattage rating on an LED lamp but can provide a higher lumen output than incandescent lamps; LEDs save more energy and are also brighter.

2.1.4 White LED

In all visible light LEDs, white LED is the most important light source for VLC and illuminations. There are two primary techniques to produce white LEDs which generate high-intensity white light [2]. Note that the *whiteness* of the light produced by these methods is essentially engineered to suit the human eye, and depending on the situation it may not always be appropriate to think of it as *white light*.

The first technique is to use a single LED chip that emits three primary colors of RGB, and then mix three colors to form white light, as shown in Fig. 2.5a. This type of LED unit is called RGB LED (sometimes also referred to as full-color LED or multicolor LED). Because this method needs electronic circuits to control the blending and diffusion of different colors, and because the individual color LEDs typically have slightly different emission patterns, this leads to variation of the color depending on direction. In addition, each LED chip in this type of LED unit is a narrow band source, that makes gaps in spectrum (Fig. 2.5b), and emission power decays exponentially with rising temperature, resulting in a substantial change in color stability. So RGB LEDs cannot provide good quality white lighting. Nonetheless, this method has many applications because of the flexibility of mixing different colors. For example, for VLC, it is an effective source to implement wavelength division multiplexing (WDM), which is an optical communication technology for transmitting large-capacity signals.

The other technique is to use a phosphor material to convert monochromatic light from a blue or UV LED to form broad-spectrum white light. This type of LED unit is called phosphor-based or phosphor-converted white LED. The luminous efficacy of phosphor-based LED unit depends on the spectral distribution of the resultant light output and the original wavelength of the single LED chip that is used to form the white LED unit. The luminous efficacy is a measure of how well a light source produces visible light. For example, the luminous efficacy of a typical yellow phosphor-based white LED unit ranges from three to five times the luminous efficacy of the original single blue LED because of the human eye's greater sensitivity to yellow than to blue. Currently, this method is the most popular method for making high-intensity white LEDs due to the simplicity of manufacturing compared to the method of multicolor LED. Figure 2.6

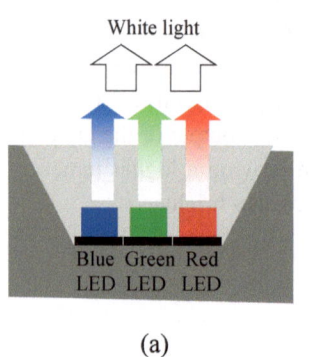 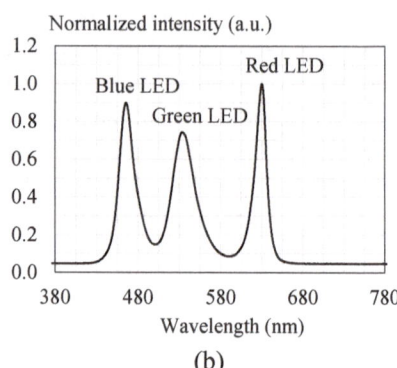

(a)   (b)

Fig. 2.5 Multicolor white LED unit formed by single RGB LED chips: **a** formation method and **b** combined spectral distribution, FWHM (full width at half maximum) spectral bandwidth is approximately 24–27 nm for each color chip

shows a yellow phosphor-based white LED unit; the formation method and the broadened optical spectrum are shown in Fig. 2.6a, b, respectively.

For a yellow phosphor-based white LED, the thickness of yellow phosphor layer and the wavelength of the blue LED chip influence its color temperature, and different color temperatures lead to different spectral distributions [3], as shown in Fig. 2.7. The color temperature is a measure of visible light color that is emitted by a light source, using a unit of measure for absolute temperature in kelvin (K). From Fig. 2.7, when the color temperature is 4000 K, it can be observed that the two peak's heights of radiant power on the wavelength spectrum are almost the same. From the perspective of communication,

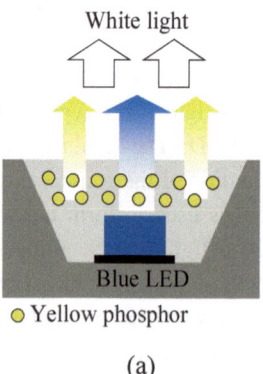 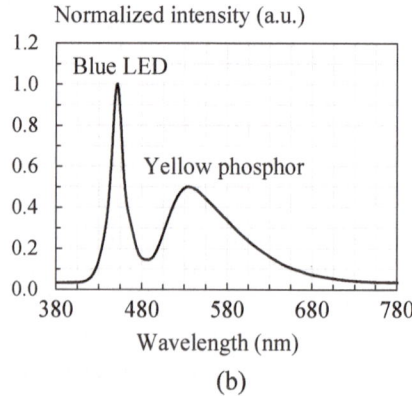

(a)   (b)

Fig. 2.6 Phosphor-based white LED unit formed by blue LED chip and yellow phosphor: **a** formation method and **b** combined spectral distribution

this characteristic implies that when phosphor-based white LED is chosen and used as a VLC light source, yellow phosphor materials that interfere with communication would have a radiance intensity equivalent to that of blue LEDs. Therefore, measures need to be taken to address the degradation of signal transmission accuracy due to color dispersion and the deterioration of response performance caused by the accumulation time of phosphor emission. In other words, it is expected that when this type's LED is chosen and used as the VLC light source, its color temperature is an important parameter that must be considered. Because only the single blue LED is employed to data transmission in VLC, the phosphor effect makes the speed and quality of the communication attenuate when the color temperature is below about 4000 K.

As a reference, Fig. 2.8 shows some practical phosphor-based white LED lamps with different color temperature. Generally, the higher the color temperature, the better for data transmission, i.e., the batter for the VLC.

White LEDs can also be made by coating near-ultraviolet LED chips with a mixture of phosphors that emit red, blue, and green light. This method is less efficient than blue LED chips with yellow phosphor, but because several phosphor layers of distinct colors are applied, the emitted spectrum is broadened, effectively raising the color rendering

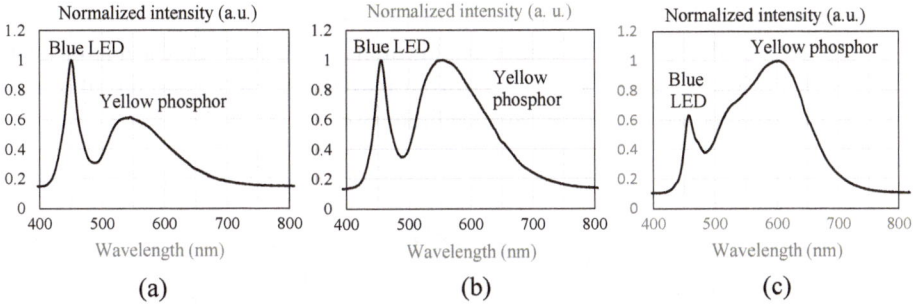

Fig. 2.7 Dependence of the color temperature on the spectral distribution in yellow phosphor-based white LED. Color temperature: **a** 5700, **b** 4000, **c** 3000 K

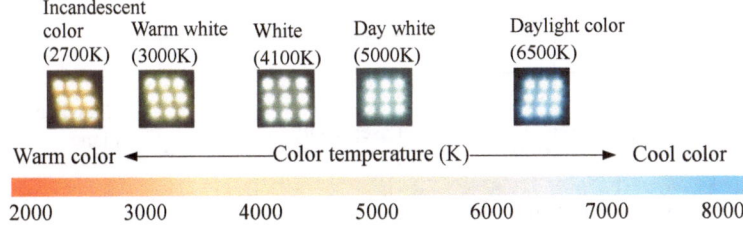

Fig. 2.8 Phosphor-based white LED lamps with different color temperature

index (CRI) value of a given LED [4]. The CRI is most useful measure of a light source's color characteristics. In general terms, CRI is a measure of a light source's ability to show object colors *realistically* or *naturally* compared to a familiar reference source, either incandescent light or daylight.

2.2   Laser Diode (LD)

2.2.1   Basic Principles

A laser diode (LD) is a device that causes laser oscillation by flowing an electric current to semiconductor. The mechanism of light emission is the same as a LED. Light is generated by flowing the forward current to a p–n junction. In forward bias operation, the p-type layer is connected with the positive terminal and the n-type layer is connected with the negative terminal, electrons enter from the n-type layer and holes from the p-type layer. When the two meet at the junction, an electron drops into a hole and light is emitted at the time, as shown in Fig. 2.9.

The difference between LDs and LEDs is that LDs make LEDs have a laser-oscillate structure i.e., a laser cavity. In the LD structure, shown in Fig. 2.9, to satisfy the laser-oscillation condition both ends of the cavity are processed and become a cleavage plane. The active layer, which is used for light amplification, has a structure capable of totally reflecting light. The laser beam is emitted from the side of the diode chip, this edge-emitting LD also known as an EEL (edge emitting laser).

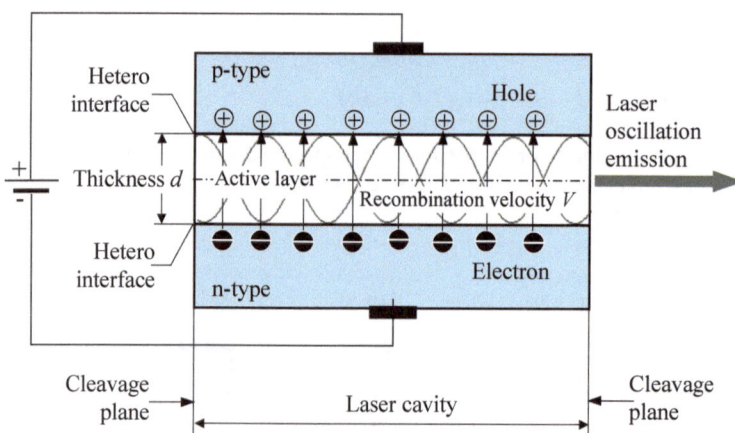

Fig. 2.9 Laser-oscillation principle of LDs

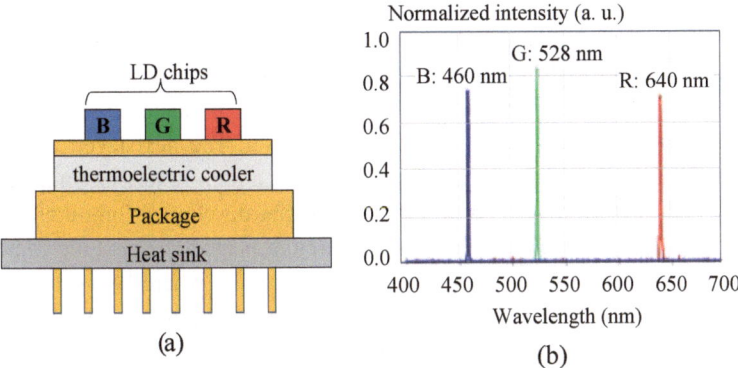

Fig. 2.10 Multicolor VLD unit formed by single RGB LD chips: **a** formation method and **b** combined spectral distribution; spectral widths are less than 5 nm

2.2.2   Visible LD

The introduction of visible LD known as a VLD, which emits coherent light in the visible spectrum (380–780 nm) can be traced back to the early 1990s. This period saw significant advancements in semiconductor materials and fabrication techniques, enabling the development of VLDs.

The semiconductor material of LDs is also the same as that of LEDs, In the early years, red laser diodes were the first to become commercially available in the visible spectrum. These red LDs typically emitted light at a wavelength of around 650–670 nm.

As technology progressed, VLDs in other colors such as green, blue, yellow, and violet started to emerge. The development of gallium nitride (GaN) and InGaN semiconductor materials played a crucial role in enabling the production of these diverse colors. Figure 2.10 shows a multicolor VLD unit formed by single RGB LD chips. The oscillation wavelengths of the RGB LD chips are 460, 528 and 640 nm, respectively; the formation method and the relatively narrow spectrum are shown in Fig. 2.10a, b, respectively [5].

Since their introduction, VLDs have continued to advance in terms of performance, efficiency, and cost-effectiveness. They have found widespread adoption in various applications, ranging from consumer electronics to scientific research and industrial systems. VLDs are also important VLC source same as visible LEDs.

2.2.3   Vertical Cavity Surface Emitting Laser (VCSEL)

Unlike conventional EELs as shown in Fig. 2.9, which emit light from the edge of the semiconductor chip, VCSELs emit light perpendicular to its top surface, as shown in

Fig. 2.11. This vertical emission enables easy coupling of light into other optical components. VCSELs can be designed to emit light across a broad range of wavelengths,
including near-infrared (NIR) and visible spectrum, making them also suitable for VLC.

Figure 2.12 shows working principles of VCSEL. It is made of several layers. The
top is a layer in electrical contact for current injection. The next layer, i.e., the second
layer, is a DBR (distributed Bragg reflector), which is a high-reflectivity mirror with 99%
reflectivity. The next third layer is an oxide layer that develops a light-emitting window
so that the light beam can be converted into a circular beam. The next the center layer in
the VCSEL is the laser cavity. It is the active gain region where lasing happens. Again
there is an oxide layer below the center layer to confine the light. And the last layer is
again a DBR, but it has more reflectivity than the top mirror of 99.9% so that lasing light
can get out from the top mirror instead of the bottom mirror.

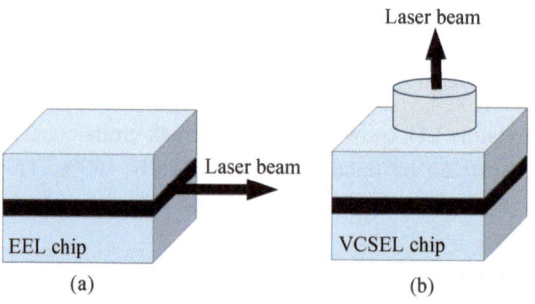

Fig. 2.11 Beam emission direction for: **a** EEL and **b** VCSEL

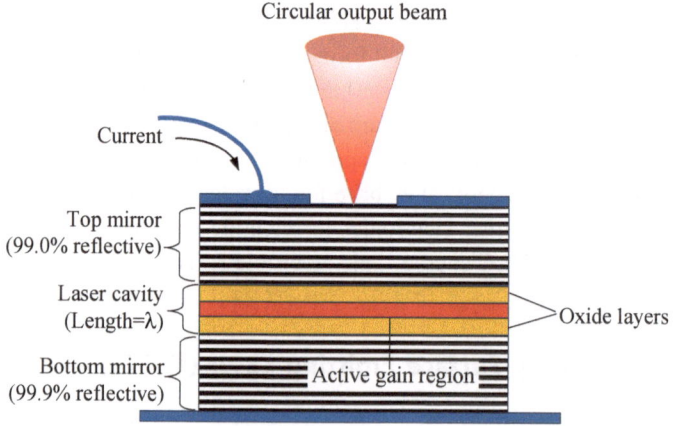

Fig. 2.12 Working principle of VCSEL

Some of the significant characteristics of the VCSEL, which makes it an important option for VLC.

- High efficiency: VCSELs are known for their high efficiency compared to other types of lasers. The vertical emission allows efficient extraction of light, and the low threshold current enables them to operate at lower power levels.
- Modulation capability: VCSELs can be modulated at high speeds, making them ideal for applications that require high-speed data transmission, such as local area networks (LANs), data centers, and so on.
- 2D array configuration: VCSELs can be fabricated in a two-dimensional array configuration, enabling them to deliver high optical power and form complex beam patterns. This feature makes them suitable for 3D VLC sensing technology.
- Low divergence: VCSELs typically have a circular output beam with low divergence, making them suitable for applications which efficient coupling of light into other optical devices.

2.3 Comparison Between LED and LD

Table 2.2 presents a comparison between LEDs and LDs [6, 7]. LEDs are incoherent sources with low cost and most LEDs emit light from a sufficiently large surface area that they are generally considered eye safe. Typical packaged LEDs emit light into semiangles (at half power) ranging from about $10°$–$30°$, making them suitable for directed transmitters for VLC. The electro-optic conversion (E/O) efficiencies of LEDs have been steadily improving over time, cutoff in 2021, typical E/O-efficiencies for commercially available LEDs ranging from about 30–60%. Potential drawbacks of LEDs include:

Table 2.2 Comparison between LEDs and LDs

Characteristics	LED	LD
Spectral width	25–100 nm	$<10^{-5}$–5 nm
Modulation bandwidth	Tens of kilohertz–tens of megahertz	Tens of kilohertz–tens of gigahertz
E/O conversion efficiency	30–60%	40–70%
Color gamut	100%	130–140%
Polarized light use	No	Yes
Directivity	$180°$	$0°$
Eye safety	Generally considered eye-safe	Must be rendered eye-safe, especially for visible LD
Cost	Low	Moderate to high

- Modulation bandwidths that are limited to tens of MHz in typical low-cost devices.
- Broad spectral widths (typically 25–100 nm), which require the use of a wide receiver optical passband, leading to poor rejection of ambient light.
- The fact that wide modulation bandwidth is usually obtained at the expense of reduced E/O efficiency.

Viewed against the incoherent and wide spectral in spatial emission of LEDs, LDs offer far superior performance. Hence, for long-distance and high-bit-rate systems, LDs are the obvious choice. However, in terms of cost, eye safety, complexity in drive circuit, reliability, temperature dependence and drive current requirements, LDs compare poorly with LEDs. LDs are much more expensive than LEDs, but offer many nearly ideal characteristics for communications:

- High E/O efficiencies of 40–70%
- Wide modulation bandwidths, which range from hundreds of MHz to more than 10 GHz.
- Very narrow spectral widths (spectral widths ranging from several nm to well below 1 nm are available).

To achieve eye safety with an LD requires that the laser output is passed through some element that destroys its spatial coherence and spreads the radiation over a sufficiently extended emission aperture and emission angle.

As laser lighting by VLDs has made great progress in recent years. this may be an important opportunity for VLC.

2.3.1 Eye Safety

The primary drawback of radiation in VL band relates to eye safety. Figure 2.13 shows ocular absorption site for radiation of different wavelengths. Radiation from visible to near infrared range (400–1400 nm) can pass through the cornea and be focused by the lens into an extremely small spot on the retina [8]. When the light energy is absorbed by the retina, it can cause permanent, although not always immediately noticeable, damage.

The coherence and low divergence angle of laser light from LDs, aided by focusing from the lens of an eye, can cause laser radiation to be concentrated into an extremely small spot on the retina, as shown in Fig. 2.14a. If the laser is sufficiently powerful, permanent damage can occur within a fraction of a second, which is faster than the blink of an eye. To achieve eye safety with an LD requires that one pass the laser output through some element that destroys its spatial coherence and spreads the radiation over a sufficiently extended emission aperture and emission angle. For example, one can employ

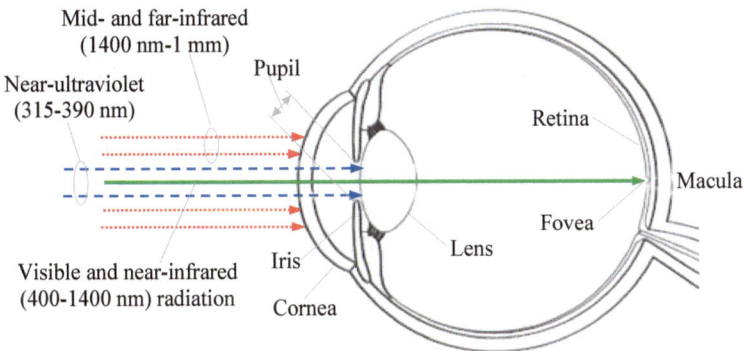

Fig. 2.13   Ocular absorption site for radiation of different wavelengths

Fig. 2.14   Spectral emission pattern of **a** LD and **b** LED

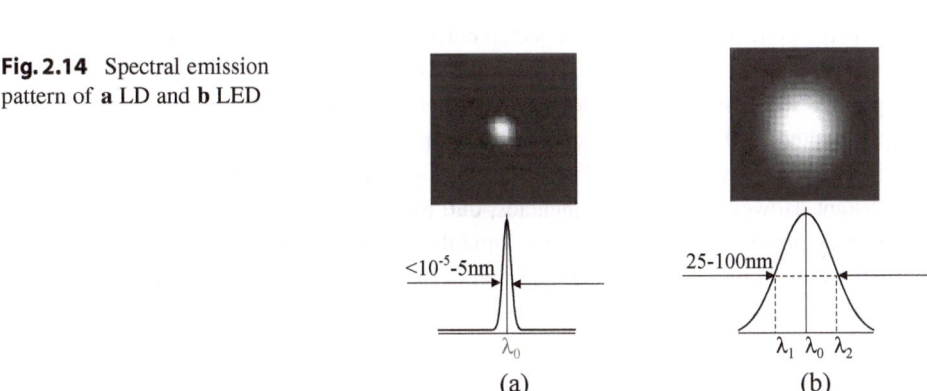

a transmissive diffuser, such as a thin plate of translucent plastic. While such diffusers can achieve efficiencies of about 70% [6].

As against the coherent and narrow spectral emission of LDs, LEDs are generally considered eye-safe because that the LED is an incoherent source, the energy is diffused over a large area before reaching human eyes, i.e., its image is dispersed in the retina, as shown in Fig. 2.14b. Also, most LEDs emit light from a sufficiently lager surface.

The eye safety of VL transmitters is governed by International Electrotechnical Commission (IEC) standards [9]. It is desirable for VL transmitters with a LD to conform to the IEC Class 1 allowable exposure limit (AEL), implying that they are safe under all foreseen circumstances of use, and require no warning labels.

2.3.2   Output and Modulation Characteristics

The output and modulation of LED/LD sources are two primary characteristics for LED/
LD-based VLC. For an injected current i of a p–n junction LED/LD, N carriers ($N = i/$
e, e is elementary electric charge) are generated. If η is the quantum efficiency, i.e., the
fraction of charges that recombine radiatively, and it describes how many electrons are
actually generated in a photoelectric process, then ηN photons will be produced. If the
bandgap is E then the emitted photons will have an energy $E = h\nu$ (h is Planck's constant
and ν is frequency of the photon). So the optical output P_o and the injected current i are
linearly related as

$$P_o = \eta NE = \eta \frac{i}{e} E. \tag{2.1}$$

Thus the optical output and the injected current are linearly related. However, at higher
levels of injected currents, harmonic distortions may occur. It is therefore important to
design electronic modulation circuits that suitably linearize the optical output and thus
keep the intermodulation distortion to a minimum. When light intensities from an LED/
LD is modulated by an electrical signal, the optical output at low modulation frequencies
is constant. However, at high frequencies, duo to the delay in the recombination process,
the output power falls off. Hence the modulation frequency f for the communication
system has a limit at which the optical output power $P_o(f)$ is reduced by -3 dB to half
of the maximum value $P_{o|\max}$. Thus, the modulation response of an LED/LD is described
by

$$P_o(f) = \frac{P_{o|\max}}{\sqrt{1 + (2\pi f \tau)^2}}, \tag{2.2}$$

where τ is the carrier lifetime. In semiconductor physics, the carrier lifetime is defined
as the average time it takes for a minority carrier to recombine. For semiconductor photo
elements such as LEDs and LDs, the carrier lifetime is used as the time constant of the
exponential decay of carriers. It includes radiative and nonradiative lifetimes, and is given
by

$$\frac{1}{\tau} = \frac{1}{\tau_r} + \frac{1}{\tau_{nr}}, \tag{2.3}$$

where τ_r and τ_{nr} represent the radiative and nonradiative lifetimes, respectively [10].

On the other hand, when a device is formed by introducing relatively low concen-
trations of impurities into the same semiconductor, the resulting junction can be called
homojunction. The normal p–n junction is formed in this manner. To achieve high efficiency, both LEDs and LDs are made with heterojunctions. In the case of heterojunctions,
τ in (2.3) is given by

Fig. 2.15 Dependence of optical power output on forward current for LEDs and LDs

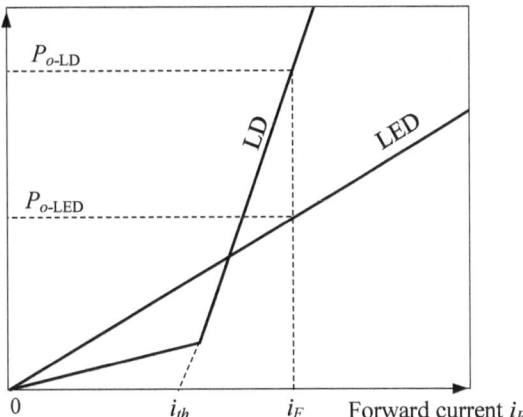

$$\frac{1}{\tau} = \frac{1}{\tau_r} + \frac{1}{\tau_{nr}} + \frac{2V}{d}, \tag{2.4}$$

where V is the recombination velocity and d is the active layer thickness (Fig. 2.9). The reorganization is a process in a semiconductor when electrons and holes disappear together and release energy.

The optical power output against the forward current for LEDs and LDs is plotted in Fig. 2.15, where i_{th} is threshold current of LD. From this figure, one can infer that an LD is a threshold device and the linear part of the response curve is small; an LED on the other hand has good linear response and is best suited for analog modulation.

References

1. K. Ando: *Principles of LED* (Ohmsha, Tokyo 2010)
2. N. Hirosaki, N. Kimura, K. Sakuma, S. Hirafune, K. Asano, D. Tanaka: White light-emitting diode lamps for lighting applications, Fujikura Tech. Rev. **109**, 1–4 (2005)
3. X. Lin: Chapter 9.2, *LED-Based Illumination-Light Communication Device* (Technical Information Institute, Tokyo 2014) pp. 609–614
4. T. Taguchi: White LED with high color rendering, ULVAC. J. **53**, 14–17 (2008)
5. T. Kumano, Y. Enya, K. Ishihara, H. Nakanishi, T. Ikegami, T. Nakamura: Ultracompact RGB laser module operating at +85°C, SEI Tech. Rev. **188**, 117–121 (2016)
6. J.M. Kahn, J.R. Barry: Wireless infrared communications, Proc. IEEE, **85**(2), 256–298 (1997)
7. K. Yamamoto: Laser lighting by visible laser diode. In Proc. 9th Light-Tech Expo Conf., Keynote Session (Tokyo, 2017) pp. 21–44 (2017)
8. B. Lee, E. Staley, L. Johnson: Laser eye safety for telecommunications systems, App. Note, www.senko.com

9. Int. Electrotech. Commission, CEI/IEC60825-1 Amendment 2: *Safety of Laser Products*, (2001)
10. A. Selvarajan, S. Kar, T. Srinivas: Chap. 4.1.1, *Optical Fiber Communication* (McGraw-Hill, New York 2002) pp. 53–58

VL Detectors

3

Compared with the light sources for VLC transmitter introduced in Chap. 2, this chapter introduces photodetectors for VLC receiver. These include the general-purpose PIN-PD (Positive-intrinsic-negative photodiode), APD (Avalanche photodiode) suitable for high-speed signal detection, and PMT (Photomultiplier tube) suitable for high-sensitivity precision detection. In addition, this chapter also introduces 2D image sensors for optical camera communications, which involving advanced metalens technology. Finally, a photon-counting detector operating in the digital domain without electronic noise is introduced.

For current VLC, even if light is used as a carrier wave, *all-optical* processing cannot be achieved (in the future, advances in optical materials or optical elements may enable an all-optical communication system). Light is only used to transmit signals. For signal detection and processing, optic-electro conversion (O/E) is still indispensable. The photodiode (PD) is used for this purpose. PDs convert the information transmitted via optical carriers to corresponding electrical signals, which are then processed further to perform the desired function. To transform optical power into electrical power, i.e., to detect the optical signal, a variety of physical effects can be utilized. The optimum choice is the use of the solid-state PD, because it can not only operate at low voltages and offer high speed and long lifetime, but also because it can be integrated to form an optoelectronic integrated circuit (OEIC) thus realizing a monolithic optical receiver of a VLC system.

Currently, two types of silicon PD are widely available for VLC: positive-intrinsic-negative (p-i-n) PD (PIN-PD) and avalanche photodiode (APD). Additionally, a PD array (also called an image sensor, IS) is often used for a VLC system requiring low speed, outdoor application, and multiple source reception.

X. Lin, *Visible Light Communications*, Synthesis Lectures on Communications,
https://doi.org/10.1007/978-3-031-64475-7_3

A photomultiplier tube (PMT) is also an effective VL-PD device used to detect very weak optical signals mainly. The versatility, extremely sensitivity, and highly speed of PMTs make them invaluable tools in VLC applications that require low-noise, high-sensitivity light detection.

A photon-counting camera (PCC) is an advanced VL imaging device that can detect individual photons of light and count them. PCCs operate in the digital domain, allowing them to provide more precise and sensitive imaging capabilities and analyze extremely weak light signals with unparalleled precision. Now, PCCs have been used in the deep space optical communications (DSOC) as the receiving device of the aircraft to receive light signals transmitted from the ground.

3.1   Positive-Intrinsic-Negative Photodiode (PIN-PD)

A silicon PIN-PD is the ideal solid-state photodetector and is employed in most receivers of VLC system at present due to its fast response, high quantum efficiency and capability to handle high electric fields and modulation frequencies.

Basic Principles

In the p-i-n structure, a long intrinsic i region is sandwiched between the p and the n layers, as shown in Fig. 3.1a. The main feature of the i region is that is has a relatively high electric field exists across it, the carriers generated within it are accelerated by the field contributing to the drift current. The thickness of the i region can be tailored so that most of the light absorption takes place within it, thus minimizing the diffusion component of the current. Hence, PIN-PDs can operate much faster than diffusion PDs. The capacitance of the junction is also small due to the large i region.

When the incident photon has energy greater than or equal to the bandgap energy E_g of the semiconductor material, the photon excites an electron from the valence band to the conduction band in the material and a photogenerated electron is generated, then leaving behind a hole (photogenerated hole) in the valence band, and an electron–hole pair is generated. This electron hole pair is called photocarrier, as shown in Fig. 3.1b. In a well-designed PIN-PD, the photocarrier generation process occurs mainly in the depletion layer of the p-i-n junction where the incident light is largely absorbed. As a result of the high electric field present in this region, the electrons and holes separate and drift in opposite direction.

Carriers generated outside but within a diffusion length on either side of the depletion region, will diffuse inward and get collected across the junction. When the photocarriers drift through the high field region, under the influence of the strong electric field generated by a reverse bias potential difference across the PIN-PD as shown in Fig. 3.1a a photocurrent I_p is induced in the load that proportional to number of incident photons, and developing an output voltage across the load resistor R_L.

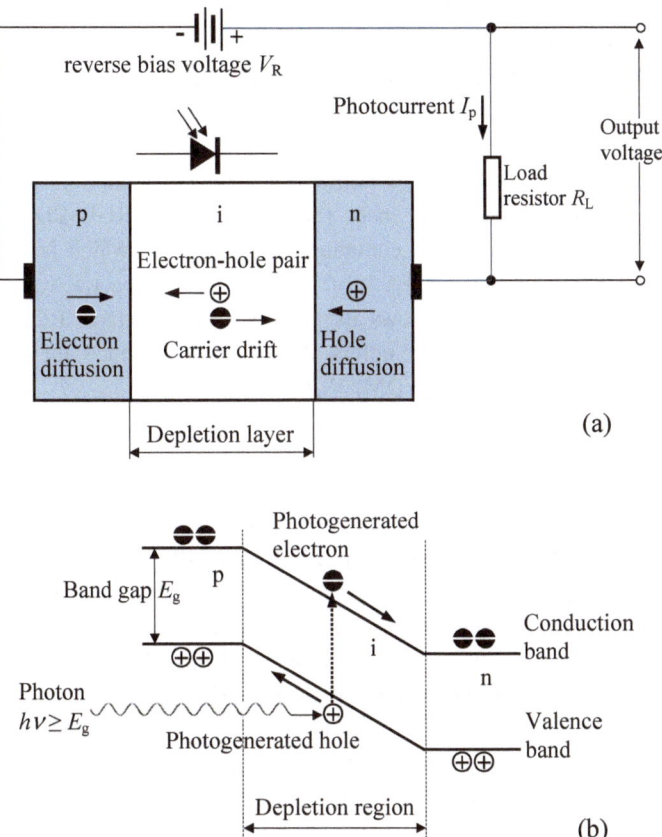

Fig. 3.1 Structure and principle of a PIN-PD

Spectral Response

When selecting the PD on the receiver side of a VLC system, in addition to the working wavelength range to match the of the LED at the transmitter side, its spectral response, i.e., the receiving sensitivity is also an important requirement to the receiver design. High receiving sensitivity can reduce the energy loss of the VLC system.

The spectral response of PDs is defined as the variation of its quantum efficiency with the wavelength of the incident light. The short wavelength cut-off of the spectral response for a PD depends on the base width and surface recombination rate. A decay of response in the region of longer wavelength appears at the absorption band edge (E_g) is called the long wavelength cut-off and is given by, $\lambda_c = (1.24/E_g)$, where λ_c is in microns and E_g is in eV.

The peak of the practical response for a PD heavily depends on the on the absorption coefficient α of the diode material. If the penetration depth $X_0 = \alpha^{-1}$ is a function of wavelength, as is the case for silicon, the peak of the spectral response can shift with a change of both the base width and the rate of recombination. Thus, in a silicon PD, this peak can be shifted in the range from 600 to 1000 nm. Figure 3.2 shows the spectral response of a high-speed response silicon PIN-PD [1]. The peak photosensitivity wavelength $\lambda_p = 760$ nm, where photosensitivity is in ampere/watt (A/W). The photosensitivity for light wavelength R (650 nm)/G (550 nm)/B (450 nm) is about 0.45/0.35/0.25 A/W.

Figure 3.3 shows a three-color sensor formed by single RGB PD chips [2]. It is three-color sensors in one package, containing PDs, each of which is sensitive to one of blue (B), green (G), or red (R). Each of PD chips has a spectral response range, as shown in Fig. 3.3a. Figure 3.3b shows its photodetection surface. RGB color sensors allow simultaneous detection of three colors of R, G, and B. Hence, they are suitable for wavelength-division-multiplexing (WDM) VLC systems.

Time Response

Fast time response i.e., the response speed is also an important requirement to design a VLC receiver. Generally, the response speed of the PD, which is used a VLC system should be higher than the overall system speed of about 10 times.

Due to the distributed generation of carriers within the diode and the different mechanisms of their transport, the time response of a PD depends on wavelength, semiconductor material, diode geometry and load impedance of the electrical circuit. The speed of a PD is basically governed by (a) the carrier drift in the depletion region and (b) the carrier diffusion in the non-depleted region (Fig. 3.1a).

The transient response of a practical PD to a square wave input optical pulse is determined by its rise time characteristics. At a sufficiently high reverse biasing voltage, the potential gradient exists throughout the PD device depth and a uniformly fast response time is achieved. Further, the response time can be improved by designing the device to

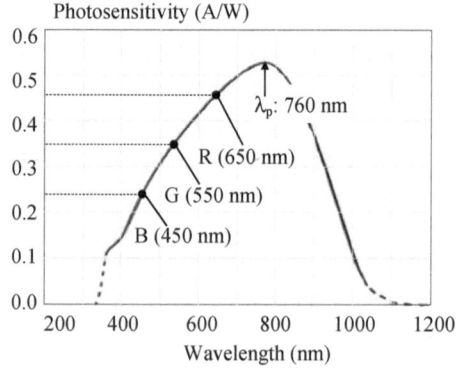

Fig. 3.2 Spectral response of the PIN-PD S10783 made in Hamamatsu Photonics

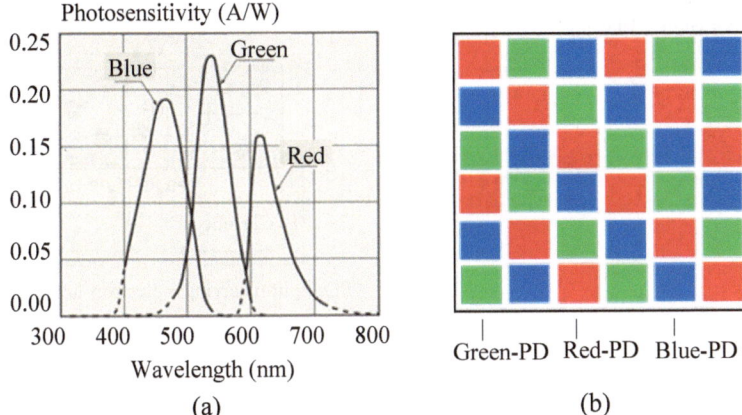

Fig. 3.3 RGB-color sensor S9702 made in Hamamatsu Photonics: **a** spectral response and **b** photodetection surface

withstand voltages which are greater than the voltage required for full depletion, creating a higher field gradient and accelerated photocarrier collection.

Comparison Between PDs and LEDs
Both PDs and LEDs are semiconductor-based p–n junction diodes that function by the interaction of light energy and electricity. The most significant difference between a PD and an LED is in the way they function. A PD converts light energy into electrical energy and operates on the principle of photoconductance (Fig. 3.4a), while an LED converts electrical energy into light and operates on the principle of electroluminescence (Fig. 3.4b).

Figure 3.5 shows a simple electronic circuit to compare the working principle of the PD and the LED. The reversing biasing does not damage PDs, because they are specially designed to work in reverse bias (although, they may also be forward biased), whereas the reverse biasing of LEDs can damage them permanently, because LEDs work in forward bias only.

In PDs, when the input light is not available, a small electrical output also present which is the leakage current is known as dark current. In order to focus the light on the p–n junction well, the structure of PDs consists of a focusing lens. There is no leakage current in case of LEDs, as these are always forward biased, there is no output of light in the absence of electrical input.

Fig. 3.4 Comparison of
principles between **a** PDs and
b LEDs

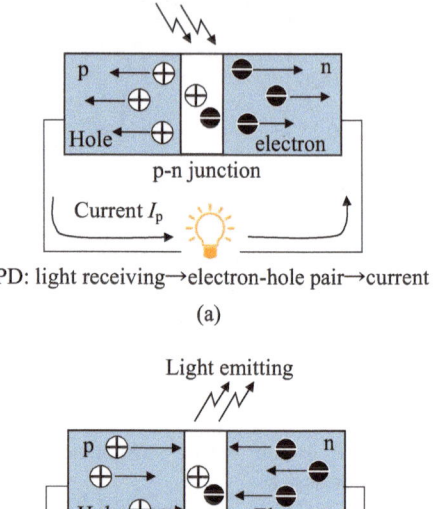

PD: light receiving→electron-hole pair→current

(a)

LED: current→electron-hole pair→light emitting

(b)

Fig. 3.5 Circuits of PDs and
LEDs

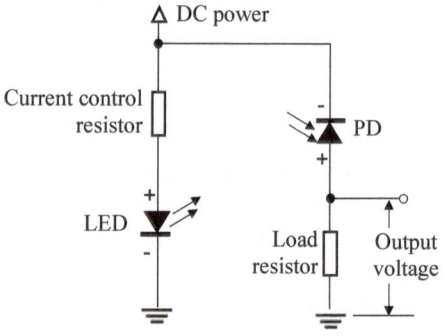

Comparison Between PDs and Solar Cells

A solar cell can also as a photodetector use to some VLC systems without high precision requirement. Its advantage is that it can provide the power required by the system while receiving light signals [3].

Solar cells are also semiconductor-based p–n junction devices that convert light energy into electric energy same as PDs, but they serve different work purposes and have distinct design, operation characteristics. PDs are primarily used for light detection, usually *output*

Table 3.1 Comparison between PDs and solar cells

	PD	Solar cell
Purpose	Detecting light signal	Generating power from light
Operation	Output photocurrent by photoconductive effect	Output voltage by photovoltaic effect
Structure	– Designing to be sensitive to light across a specific range of wavelengths – Optimizing for high-speed response and can quickly detect changes in light intensity	– Optimizing to efficiently photoelectric conversion across a broader range of wavelengths – Designing to maximize energy conversion efficiency rather than high-speed response
Biasing	Operating in bias to increase its sensitivity to light	Operating under zero bias or at a slight forward bias

current by photoconductive effect, and can respond quickly to changes in light intensity. Solar cells, on the other hand, are used for converting light into electricity, usually *output voltage* by photovoltaic effect, also known as a photovoltaic cell, and are optimized for high energy conversion efficiency. Table 3.1 list a comparison between PDs and solar cells.

3.2   Avalanche Photodiode (APD)

An APD is essentially p-i-n device, where the incident photons generate primary electrons and holes. APDs consist of p^+ and n^+ region which is heavily doped. Also, there are another two lightly doped regions i.e., intrinsic region i (p^-) and p region, the intrinsic region is also called photoabsorption region, as shown in Fig. 3.6. Here is in the case of short wavelength (i.e., visible light), which contrary to the near-infrared APD, photon hv is structured to enter from the p^+ layer [4].

In APDs, relatively high (around 20 V) reversed biased or reversed voltages are applied to the photodiode which accelerates the electrons with high energy. These electrons and holes strike the neutral atoms and separate the other bonded electrons and holes. This is known as a secondary process causing avalanche actions. Thus, one photon eventually generates multiple charge carriers. This means photodiode internally amplifies the photocurrent.

By controlling the reverse bias voltage V_R, an arbitrary current multiplication factor M can be obtained, and M is given by Miller's empirical formula

$$M = \frac{1}{1 - (V_R/V_B)^m},$$

(3.1)

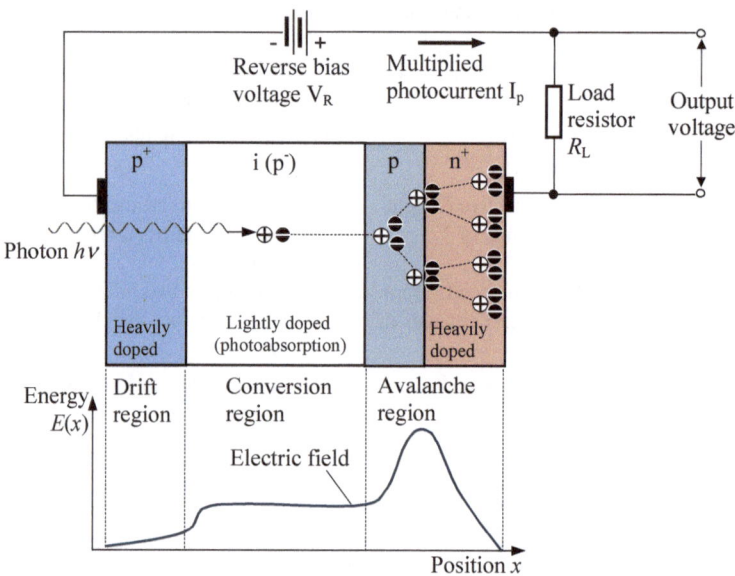

Fig. 3.6 Structure and principle of an APD

where m is a constant determined by the APD, and $V_R < V_B$. Normally, the operation region of an APD is $M = 30–100$ [5]. However, it is important to note that since the reverse breakdown voltage V_B of the APD has a large temperature dependency, the multiplication factor also strongly depends on the temperature.

The application and performance of APD depend on many parameters. Two of the most important factors are:

Quantum efficiency: Which indicates how well incident optical photons are absorbed and then used to generate primary charge carriers. APDs have a quantum efficiency of more than one (10–100) which is M times larger than the normal PIN Photodiode.

The total leakage current: Which is the sum of the dark current, photocurrent, and noise. APDs are favored in optical receivers with direct detection when there is little ambient-induced shot noise (shot noise is the inherent electronic noise that occurs in photoelectric devices, originating from the discrete nature of electric charge and associated with the particle nature of light), because its internal gain helps overcome the thermal noise from its electronic circuit, increasing the receiver SNR. APD-based receivers can lead to good VL link performance when ambient light is weak. When ambient-induced shot noise is dominant, however, use of an APD results in a net decrease in SNR, because the random nature of an APD's internal gain increases the variance of the shot noise by a factor greater than the signal gain.

For a receiver of VLC, despite the higher dark current and the excess noise due to multiplication problems in APDs, in general they can provide 3–5 dB higher sensitivity than PIN-PDs above 100 Mbps. In the case of lower bit rates, low leakage currents of PIN-PDs result in better sensitivity. Additional drawbacks of APDs include their high cost, requirement for high bias, and their temperature-dependent gain. Therefore PIN-PDs are preferred in most VLC systems at present.

3.3   Photomultiplier Tube (PMT)

A PMT is a highly sensitive photodetector used to detect and amplify very weak optical signals. It can be used in VLC receiver where high sensitivity and low noise detection of light are required. Figure 3.7 shows the schematic construction and principle of a PMT [6].

The PMT is a vacuum tube consisting of an entrance window, a photocathode, focusing electrodes, an electron multiplier (dynodes), and an anode sealed usually into an evacuated glass tube. Light entering a PMT produces an output signal through the following processes:

1. Incident light (i.e., photon) passes through the *entrance window* and strike the surface of the *photocathode*, and then excites electrons (primary electrons) in the photocathode so that photoelectrons are emitted into the vacuum via the external photoelectric effect. Photoelectric conversion is broadly classified into *external photoelectric effect* (EPE) by which photoelectrons are emitted into the vacuum from a material and *internal photoelectric effect* (IPE) by which photoelectrons are excited into the conduction band

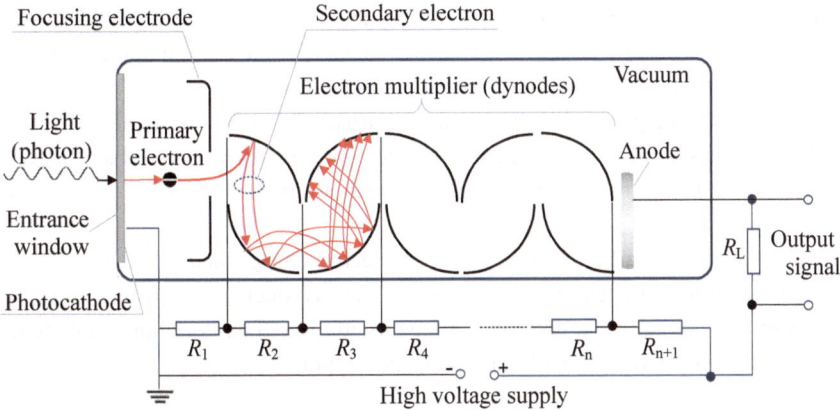

Fig. 3.7 Construction and principle of a PMT

of a material. The photocathode with the EPE in the front end of the PMT, it is one of key components in the PWT. It is the typically made of a material that has a high quantum efficiency, meaning it efficiently converts incoming photons into electrons.

2. Photoelectrons are accelerated and focused by the *focusing electrode* and impinge on the first dynode where they are multiplied by means of secondary electron emission.
3. The secondary electron emission is repeated at each dynode (electron-multiplier stages) in the *electron multiplier*. The electron multiplier is the heart of the PMT. Each dynode is at a higher positive voltage than the previous one, causing secondary emission of electrons as the primary electrons strike the surface. This process multiplies the number of electrons, resulting in significant amplification of the original signal.
4. A cluster of secondary electrons emitted from the last dynode are finally multiplied up to 10^6 to 10^7 times and are extracted from the *anode*. The anode at the end of the electron multiplier chain, the multiplied electrons are collected and form the output signal in here. The anode is kept at a high positive voltage to attract and collect the electrons, generating a measurable current or voltage pulse.

The versatility, sensitivity, and speed of PMTs make them invaluable tools for VLC systems that require highly sensitive light detection and transmission.

3.4   Image Sensor (IS)

VLC using ISs (cameras) is called image sensor communication (ISC) or optical camera communication (OCC). An IS is essentially a PD-array device that consists of a PD array chip for light detection and an integrated circuit (IC) chip for signal processing [7].

Basic Principles
The basic principle of ISs is to use the properties of semiconductors to store electrons and quantify the amount of electrons. The main devices are CCD (charge-coupled device) and CMOS (complementary metal oxide semiconductor).

In a CCD, the charge is converted to an electrical signal by means of a CCD transfer path. This process allows for high-quality image capture with low noise but may consume more power, as shown in Fig. 3.8a. The signal readout method of CCDs as the following:

1. Receives light with a PD, converts it to an electric charge, and stores it,
2. Transfer charge to vertical transmission line, i.e., vertical shift register,
3. Transfer the charge to the horizontal transmission line, i.e., horizontal shift register,
4. Convert charge to voltage with amplifier and output.

In a CMOS image sensor, it includes a CMOS transistor switch and a photodiode for every pixel, permitting the pixel signals to be improved separately. By operating these

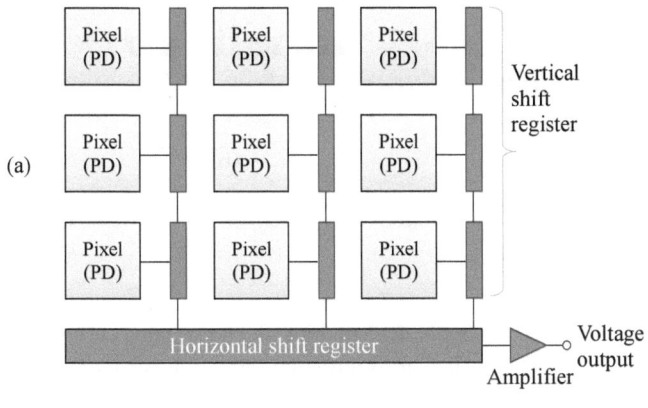

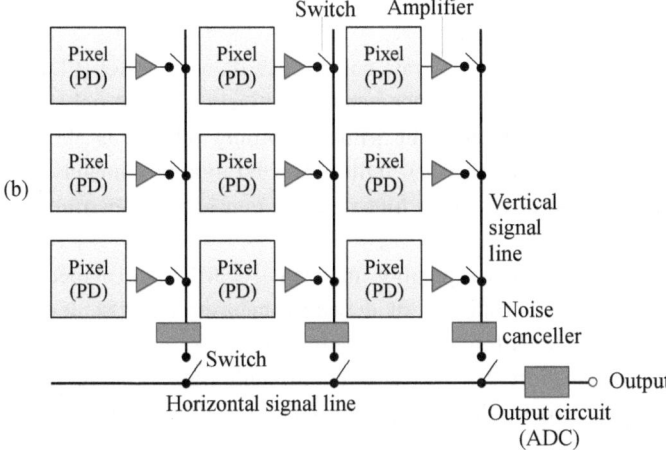

Fig. 3.8 Structure and principle of: **a** a CCD and **b** a CMOS

switches, the signals can be allowed straight and in sequence with high-speed compare with a CMOS sensor, as shown in Fig. 3.8b. Including an amplifier for every pixel can also provide another benefit: it decreases the noise that arises as understanding the electrical signals which are changed from captured light. The signal readout method of CCDs as the following:

1. Receives light with a PD, converts it to an electric charge, and stores it,
2. Convert charge to voltage with amplifier in each pixel,
3. Transfer voltage to vertical signal line by turning on/off switch for each pixel,
4. Eliminate noise with the noise canceller in each row,
5. The voltage is transferred to the horizontal signal line and output by turning on/off the switch for each column.

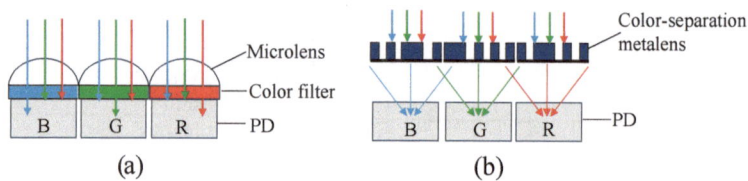

Fig. 3.9 Acquisition method of color information by an IS: **a** using color filters, **b** using color-separation metalenses

Since the CMOS is driven by a single device, it has the advantages of low power consumption and high processing speed. It is also attracting attention for its lower manufacturing cost than the CCD.

Acquisition of Color Information

In current IS, it is common to integrate color filters on a per-pixel (PD) basis to acquire color information. Color filters, on one hand, selectively transmit only the desired wavelength band (color) of incident light while absorbing others, as shown in Fig. 3.9a. Consequently, the amount of received light is limited to approximately 30% of the incident light intensity. Therefore, color filters can be considered a barrier to improving the sensitivity of IS.

On the other hand, a color-separation metalens (CSML) is designed using the wavelength dispersion control properties of dielectric metasurfaces [8]. It has the ability to focus incident light onto different PDs based on its color, as shown in Fig. 3.9b. Using CSMLs, color information can be acquired through light separation instead of light absorption, thus maximizing the amount of light received by the IS.

Spectral Response

Figure 3.10a, b shows the spectral response curves of mono and RGB color sensor using same CMOS family, respectively [9]. The majority of machine vision color cameras have IR cut filters installed to block near-IR wavelengths. This removes IR noise and color cross-over from the image, best matching how the human eye interprets color. However, in a number of applications it can be beneficial to image without the IR cut filter. Whether or not an IR cut filter is installed a color sensor will never be as sensitive as the mono sensor.

ISC Systems

In general, a PD is used as a reception device of the VL signal for a VLC system. However, in a vehicular environment when cars are moving, the PD may receive not only the desired VL signal sources but also various noise sources such as sunlight, streetlights, and other ambient lights and LED transmission sources. As results, all background lights

Fig. 3.10 Spectral response using same CMOS sensor family: **a** mono sensor and **b** RGB color sensor that on IR cut filter

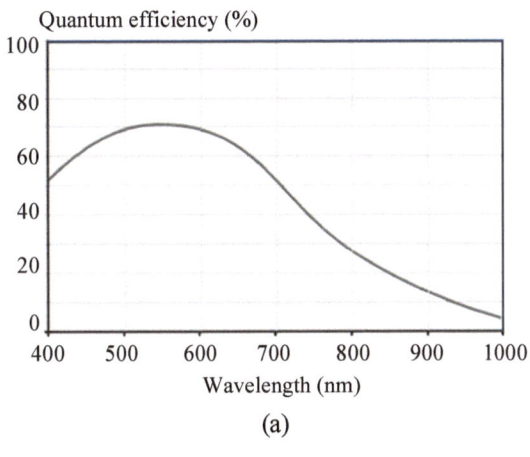

(a)

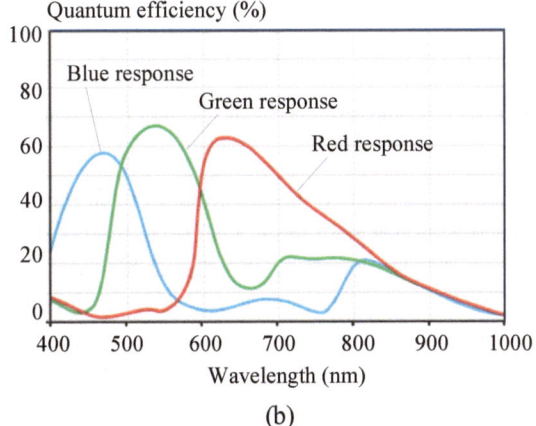

(b)

in the field of view (FOV) are summed up as noise, and the received SNR becomes severely low.

In an IS-based VLC (i.e., ISC) system, the transmitted data are received by extracting the luminance corresponding to the VLC transmitter from the captured image. Its receiver can spatially separate multiple or noise sources due to IS's parallel effect of its array structure, simultaneously demodulate multiple sources, remove noise sources such as sunlight, and detect VL signals even in daytime or outdoor environments. For example, as shown in Fig. 3.11, single PD mixes and receives light from multiple sources. That is, it is affected by noises from background light or others signal sources. However, in the ISC system, multiple sources can be simultaneously received by passing through a lens, and parallel communication can be performed when multiple sources are separated by the signal processor [10].

In addition, in contrast to the PD-based VLC system, the ISC system does not require precise adjustment of the optical axis between the transmitter and the receiver. However,

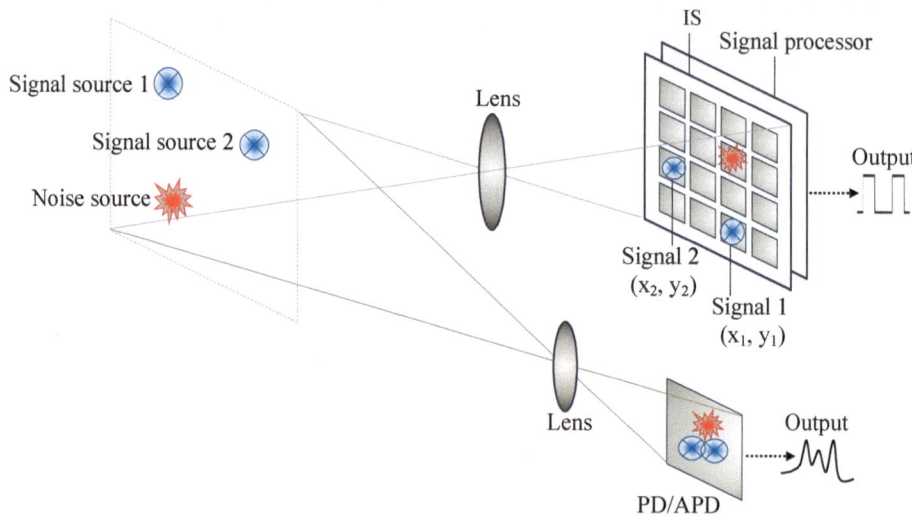

Fig. 3.11 Comparison between PD and IS for removing noise sources

since the IS needs to process images from multiple sources in parallel, image processing speed is slower than the PD. But day by day different kinds of IS sensors is available in the market by improving the size, speed, resolution and light sensitivity of ISs. In 2010, a CMOS sensor dedicated to ISC was developed by Shizuoka Univ. achieving high-speed communication of 10 Mbps at 70 m outdoors [11]. Comparisons of characteristics for PD- and IS-based VLC system are tabulated in Table 3.2 [12].

Table 3.2 Comparison between PD-based and IS-based VLC system

	IS-based VLC	PD-based VLC
Object	Temporal change of 2D images	Temporal change of signals
Coordinate	2D space $\times$ time: (x, y, t)	Time: (t)
Response	Unit impulse at specific coordinates: $\delta(x, y, t)$	Unit impulse: $\delta(t)$
Distance	Long distance possible (depending on optical system)	Relatively short (about 10 m)
Channel	Multiple (= pixels)	Single
Speed	Relatively slow	High speed possible
Spatial resolution	Pixel-by-pixel	Know the direction
Robustness	Strong (no interference due to spatial separation)	Weak

3.5 Photon-Counting Detector (PCD)

A PCD is an advanced light-signal-detection sensor that can detect individual photons of light and count them. Unlike traditional PD that storing charges and then as analog current or voltage to output, PCD operate in the digital domain, allowing them to provide more precise and sensitive detecting capabilities. Figure 3.12 shows signal-detection different method for PD and PCD [13].

PCDs are particularly useful in DSOC (deep space optical communication) system where ultra-low light levels need to be detected and where high spatiotemporal resolution is essential [14]. The key features of PCD as the following:

1. Counting Individual Photons: PCDs treat light signals not as continuous analog signals but as discrete counts of individual photons. Photons are the smallest units of light, and by accurately counting their numbers, it is possible to assess the intensity and temporal characteristics of the light signal.
2. Discrete Digital Signals: In PCDs, light is detected as discrete digital signals representing the number of photons, such as one, two, or three photons. This approach eliminates the influence of electrical noise typically associated with analog signals, ensuring a high SNR.
3. High-Precision Detection: PCDs excel at detecting weak light signals with high precision. This sensitivity is achieved because it directly counts the number of photons present in the signal.

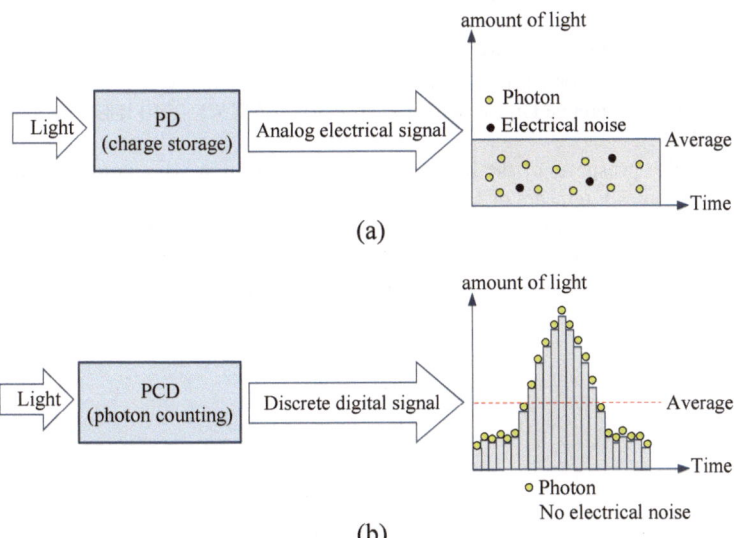

Fig. 3.12 Comparison of the signal-detection method between **a** PD and **b** PCS

4. Analysis of Time Distribution: PCD allows for high-precision measurement of the times at which photons are detected, i.e., *timestamping* capability. This capability is particularly useful for analyzing the temporal distribution of signals, aiding research in optical phenomena and dynamics.

PCDs play a crucial role in many fields where highly sensitive and precise detection of light signals is required, making them an important receiving sensor in VLC.

References

1. Hamamatsu Photonics K.K: *Si photodiodes: 100 MHz to less than 1 GHz, S10783* (Hamamatsu, Hamamatsu 2022) p. 16
2. Hamamatsu Photonics K.K: *Si photodiodes: RGB color sensors, S9702* (Hamamatsu, Hamamatsu 2022) p. 33
3. X. Lin: Optical wireless ubiquitous information service using LED lighting, Mon. Disp. 18(10), 46–52 (2012)
4. Hamamatsu Photonics: *Si APD, MPPC* (Hamamatsu, Hamamatsu 2017)
5. J. Sakai: *Optical Communication Engineering* (Kyoritsu Shuppan, Tokyo 2000)
6. Hamamatsu Photonics K.K: *Photomultiplier—Basics and Applications—*(Hamamatsu Photonics K.K., Hamamatsu 2017)
7. Hamamatsu Photonics: *Image Sensors—Selection guide—*(Hamamatsu, Hamamatsu 2018)
8. M. Miyata: High-sensitivity, multidimensional imaging based on dielectric metasurfaces, Optronics (9), 85–91 (2023)
9. Image courtesy of LUCID Vision Labs
10. H. Tanaka, M. Bandai, T. Watanabe: Fundamental discussion and experiments of visible light communication using two-dimensional code, WiNF2010 (8), 165–170 (2010)
11. S. Itoh, I. Takai, Md. S. Z. Sarker, M. Hamai, K. Yasutomi, M. Andoh, and S. Kawahito: A CMOS image sensor for 10Mb/s 70m-range LED-based spatial optical communication, Solid-State Circuits Conference Digest of Technical Papers (ISSCC), 2010 IEEE International, 402–403 (2010)
12. N. Iizuka: The prospects of the image sensor communication, The journal of the Institute of Image Electronics Engineers of Japan, 40(1), 251–256 (2011)
13. Canon: *The development of the world's first 1-million-pixel SPAD image sensor* NEWS RELEASE 2020/06/24
14. https://www.nasa.gov/mission_pages/tdm/dsoc/index.html

VLC System Designs

4

The VLC system consists of the light sources introduced in Chap. 2, the photodetectors introduced in Chap. 3, and the spatial channel. The design of the VLC system mainly includes link configurations, system construction methods, and spatial channel analysis. This chapter focuses on the technical content of these aspects, it consists of various optical wireless links, the commonly used IM/DD (intensity modulation with direct detection) technology for constructing an optical wireless communication system, and it provides a detailed analysis for various noises characteristics in the spatial channel, theoretical upper limit on the transmission rate (i.e., Shannon's theorem) in the spatial channel as well as the receiver SNR (signal-to-noise ratio). Finally, the multipath effect generated by reflections and plural sources during indoor ray transmissions is discussed.

VLC system designs general consist of link configurations, system constructions, and spatial channel analysis.

4.1 VLC Links

VLC links may employ various designs, and it is convenient to classify them according to two criteria [1], as shown in Fig. 4.1. The first criterion is the degree of directionality of the transmitter and receiver (three rows in Fig. 4.1). *Directed links* employ directional transmitters and receiver, which must be aimed in order to establish a link, while *nondirected links* employ wide-angle transmitters and receivers, alleviating the need for such pointing. Directed link design maximizes power efficiency, since it minimizes path loss and reception of ambient light noise. On the other hand, nondirected links may be more convenient to use, particularly for mobile terminals, since they do not require aiming of

© The Author(s), under exclusive license to Springer Nature Switzerland AG 2025   47
X. Lin, *Visible Light Communications*, Synthesis Lectures on Communications,
https://doi.org/10.1007/978-3-031-64475-7_4

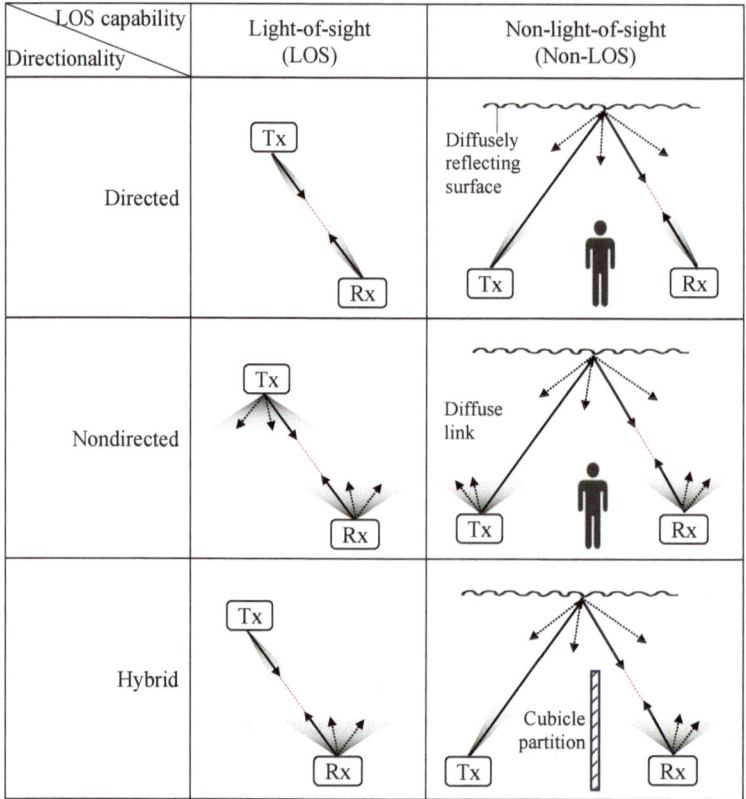

LOS capability / Directionality	Light-of-sight (LOS)	Non-light-of-sight (Non-LOS)
Directed		Diffusely reflecting surface
Nondirected		Diffuse link
Hybrid		Cubicle partition

Fig. 4.1 Link configurations for VLC systems. Tx: transmitter and Rx: receiver

the transmitter or receiver. It is also possible to establish *hybrid links*, which combine transmitters and receiver having different degrees of directionality.

The second classification criterion relates to whether the link relies upon the existence of an uninterrupted line-of sight (LOS) path between the transmitter and receiver, i.e., LOS refers to the ability to see the transmitter from the receiver (two columns in Fig. 4.1). *LOS links* rely upon such a path, while *non-LOS* links generally rely upon reflection of the light from some other diffusely reflecting surface. LOS link design maximizes power efficiency and minimizes multipath distortion. Non-LOS link design increases link robustness and ease of use, allowing the link to operate even when barriers, such as people or cubicle partitions, stand between the transmitter and receiver. The greatest robustness and ease of use are achieved by the nondirected-non-LOS link design, which is often referred to as a *diffuse* link (row 2, column 2 in Fig. 4.1). Incoherent LED-based VLC systems are suitable for the nondirected or the diffuse link.

4.2   VLC Systems

4.2.1   System Configurations

The VLC system is schematically depicted in Fig. 4.2. It consists of a transmitter, a receiver, and spatial channel. The major difference between VLCs and fiber optic communications (FOCs) is the propagation channel. In FOCs, light waves propagating in waveguide are used as propagation media, so they are also called optical cable communications or optical wire communications. In comparison, VLCs use directly spatial light waves as their propagation media, hence they are also called free-space optic (FSO) communications or optical wireless communications.

LEDs or LDs are used as light-emitting elements for the VLC transmitter. The difference between LED and LD in the interference properties of the light, which is an important characteristic quantity as a spatial wave. LEDs are incoherent sources, they suitable for the nondirected or the diffuse link, so usually LED-based VLC systems can without the need to be equipped with an optical antenna. on the other hand, LDs are coherent sources with accurate directionality, they generally are used for directed links and optical antenna is needed. The light from LEDs or LDs is modulated by the electrical signal to be sent. And then, the modulated light signal is amplificated to propagate.

At the receiving side, PDs or APDs are used usually as photodetectors. Of course, PMTs, ISs, and PCDs can also are chosen as photodetectors according to different usage requirements and purposes (see Chap. 3). The light signal from transmitter via spatial optical channel is detected and converted become an electric signal by the photodetector, and then, the electric signal is demodulated and amplificated to output at the receiving end.

The purpose of any communication system is to accurately extract the signals sent by the transmitter. Theoretically, the greater the carrier power, the higher the accuracy of the received signal. However, a good communication system should also have low power consumption, that is, having as minimal carrier power as possible to transmit data. For

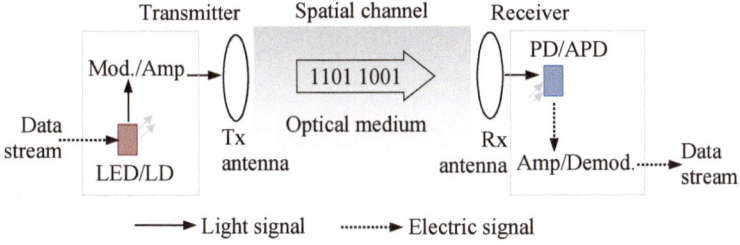

Fig. 4.2 The VLC system. Tx: transmitting, Rx: receiving, Mod.: modulation, Demod.: demodulation, Amp: amplification

VLC systems (i.e., FSO systems), the key factors affecting the performance of received power are path losses and optical antennas. The average received light power P_r of VLC systems is given by [2]

$$P_r = P_t \cdot G_t \cdot L_f \cdot L_p \cdot G_r, \tag{4.1}$$

where P_t is the transmitted average power, G_t and G_r are the optical antenna gains for transmitting and receiving, L_f and L_p are the free-space path and propagation path loss, respectively.

Optical antenna gain is the degree of aperture of the beam compared to uniform radiation, and given by

$$G = \eta(\pi D/\lambda)^2, \tag{4.2}$$

where D is the diameter of the antenna and η is the aperture efficiency, which the percentage of effective use of opening area ($= \pi D^2/4$), λ is light wavelength use to propagation medium. For the nondirected-link system without optical antenna, $G = 1$ ($= G_t = G_r$). The free-space path loss L_f is given by

$$L_f = (\lambda/4\pi d)^2, \tag{4.3}$$

where d is the distance between transmitter and receiver. The propagation path loss L_p due to fog, rain, obstacles, ambient stray light, circuit thermal noises, etc., will be discussed in detail in Sect. 4.3.

As an example, Fig. 4.3 depicts a VLC point-to-point duplex system composed by two identical transceivers that exchange data by two beams [3].

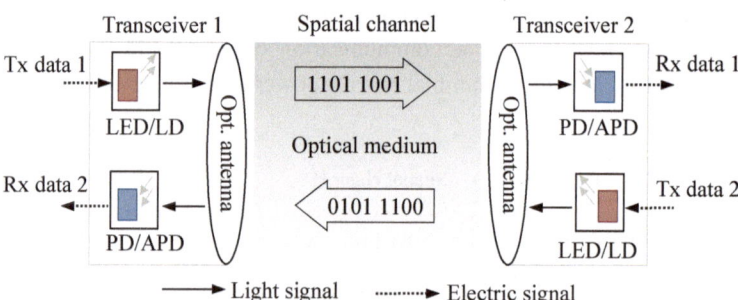

Fig. 4.3 VLC point-to-point duplex system. Tx: transmitting, Rx: receiving, Opt.: optical

4.2.2   IM/DD Techniques

For modulation and detection of VLC systems, as described in Sect. 1.2.2, the most viable modulation at present is *intensity modulation* (IM), in which the desired waveform is modulation onto the instantaneous power of the carrier. And the most simple and cost-effective signal-detection technique is *direct detection* (DD), in which a photodetector produces a current proportional to the received instantaneous power, i.e., proportional to the square of the received electric field [1, 4]. The IM/DD technique takes advantage of the characteristics of light-emitting and receiving elements.

A physical model of a baseband IM/DD-based VLC method is as shown in Fig. 4.4. It consists of an IM module, spatial channel that includes noises from the ambient environment, and a DD module that includes noises from O/E conversion and the electrical circuit.

In IM module, the light intensity from a light source (often an LED or an LD), is modulated to encode the binary data. The modulation can be achieved using various techniques, such as on–off keying (OOK), pulse position modulation (PPM), and so on, where the intensity of the light signal is directly varied based on the input data. The baseband intensity modulation (BIM) techniques for VLC systems will be discussed in detail in Sects. 5.1 and 5.2.

The modulated light signal is then transmitted through the spatial channel, which carries the signal to the receiver. It is important to note that many environments, i.e., spatial channels contain intense VL radiation arising from sunlight, skylight, incandescent and fluorescent lamps, and other sources. The resulting DC photocurrent causes shot noise, which is a dominant noise source in typical VLC receivers. Also, in the case of outdoor environments, as described in Sect. 4.2.1, VLC systems also have to deal with different atmospheric conditions that could dramatically impair the system performances. Fag, low

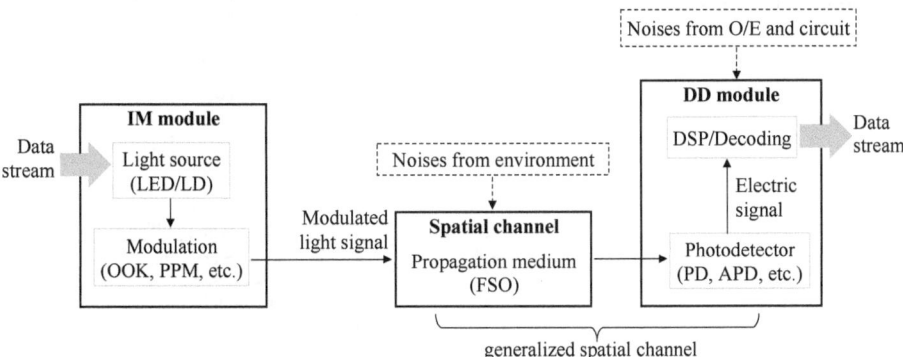

Fig. 4.4 Physical model of baseband IM/DD-based VLC method

altitude clouds, heavy rain, snow, dust, and haze increase the power loss and deteriorate the SNR. The noise analysis of VLC systems will be discussed in detail in Sect. 4.3.

In DD module, at the receiving end, a photodetector (often a PD or an APD) is used to detect the modulated light signal. The photodetector converts the received optical signal back into an electrical signal. The electric signal obtained from the photodetector is then processed by using digital signal processing (DSP) techniques to extract the transmitted data. This involves various signal processing techniques such as digital low-pass filtering, equalization, 0 or 1 recognition, and error correction coding. And then, converting the electrical signal back to its original binary form, which represents the encoded data. Finally, the processed binary signal is decoded to retrieve the original data that was transmitted. Dominant noise sources in DD module are the shot noise from O/E conversion of photodetector as all as the thermal noises from electrical circuit around photodetector, they will be discussed in detail in Sect. 4.3.

On the other hand, as a comparison of the DD method, optical heterodyne detection (OHD), also called coherent detection (CD) is also an effective method to detect signals when using the coherent source such as LDs [5, 6]. Figure 4.5 shows a comparison between methods of the DD and the OHD.

Optical heterodyne detection is a type of coherent detection that utilizes the principle of beating two light signals together to produce an electrical signal at the difference frequency. In this method, as shown in Fig. 4.5b, the incident light signal with frequency f_s and a local oscillator signal of slightly different frequency f_L are mixed on the light-receiving surface of the photodetector by using a beam combiner/splitter to form a coherent light field. Incident signal beam and local oscillator beam are spatially coherent, this means not only that their intensity profiles overlap, but also that their wavefronts have the same curvature on the detector, so that the interference conditions are uniform

Fig. 4.5 Comparison between methods of: **a** the DD and **b** the OHD

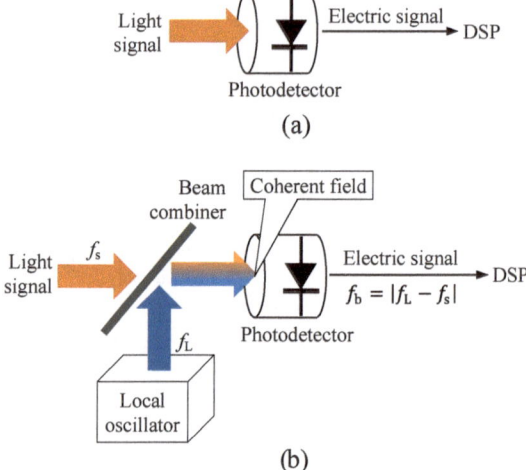

over the full detector area. Addition, the local oscillator can be phase-locked to ensure coherence.

The output from the photodetector is an electric signal with a difference frequency, i.e., beat frequency f_b. The beat frequency is equal to the complete value of the alteration in the frequency of the two light waves. Since the amplitude of the output signal is the product of the two beams, and the resulting photocurrent by photodetector is proportional to the total light intensity it is equivalent to improving the SNR of the system. This means by using the OHD method with a strong local oscillator, the heterodyne signal resulting from a weak input signal can be much more powerful than for DD method. In that sense, the OHD provides a signal gain, offers better performance and higher sensitivity, although there is no optical amplification involved. But the OHD is more complex and expensive than the DD method.

Figure 4.6 is an example of LED-based IM/DD method. It is important to note that while IM/DD is a straightforward technique, it is susceptible to various impairments, such as dispersion, attenuation, and noise, which can limit its application for long-distance and high-speed communication. However, advancements in optical amplification and signal processing have improved its performance, making it suitable for a wide range of applications.

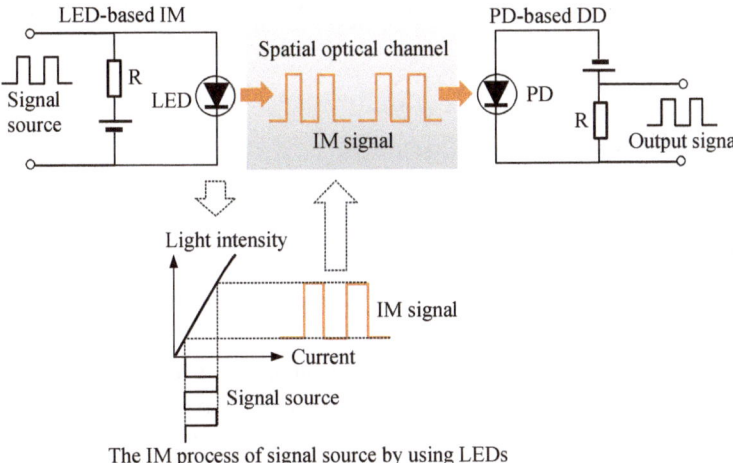

The IM process of signal source by using LEDs

Fig. 4.6 IM/DD method by using the LED source

4.3   Spatial Channel

From the perspective of noise, the spatial channel of the VLC system should also include the electronic circuits of the receiver, which can be considered as a generalized spatial channel, as shown in Fig. 4.4. In order to analyze the noises, VLC systems can be divided into two categories, which are outdoor and indoor systems according to the working environment. For VLC systems that work in natural environments such as outdoors, dominant noise sources are various particles in the atmosphere, ocean, deep space, and so on. Light propagation characteristics that include attenuations, losses, etc., in the atmosphere and seawater in detail can be found in Refs. [7, 8].

This section mainly analyzes the noise sources from indoor-environment VLC systems and the resulting receiver SNR. In fact, most VLC systems work in indoor. Assuming that the spatial channel as shown in Fig. 4.4 is an indoor environment and distortionless, i.e., it is evenly distributed for all work frequencies of interest. Hence, the noise output of the receiver is a Gaussian noise pattern [9], having a total variance (ampere squared: A^2) that is the sum of contributions from shot and thermal noise

$$\sigma_{\text{total}}^2 = \sigma_{\text{shot}}^2 + \sigma_{\text{thermal}}^2. \tag{4.4}$$

In particular, when the thermal noise of a VLC system is overwhelmingly larger than shot noise, it is called *thermal-noise-limit state*. Conversely, it is called a *shot-noise-limit state*.

4.3.1   Shot Noises

When a photodetector such as PD/APD receives an instantaneous optical power $p(t)$, it produces an instantaneous photocurrent

$$i(t) = i_{\text{r}} + i_{\text{d}} = Rp(t), \tag{4.5}$$

where, i_{r} (ampere: A) is received signal-light current, i_{d} (A) is dark current due to *ambient light*, and R (ampere/watt: A/W) is receiving sensitivity of receiver. When using a white LED as the sending source, R can be given by an integration of the wavelength distribution of the white LED $W_{\text{LED}}(\lambda)$ and the PD $W_{\text{PD}}(\lambda)$

$$R = \frac{1}{n} \int_{\lambda_1}^{\lambda_n} W_{\text{LED}}(\lambda) W_{\text{PD}}(\lambda) \mathrm{d}\lambda, \tag{4.6}$$

where, n is the sampling number of wavelength. For VLC systems, the integration range from λ_1 to λ_n is generally 380–780 nm, as shown in Fig. 4.7.

Fig. 4.7 Example of receiving sensitivity of a white-LED-based system

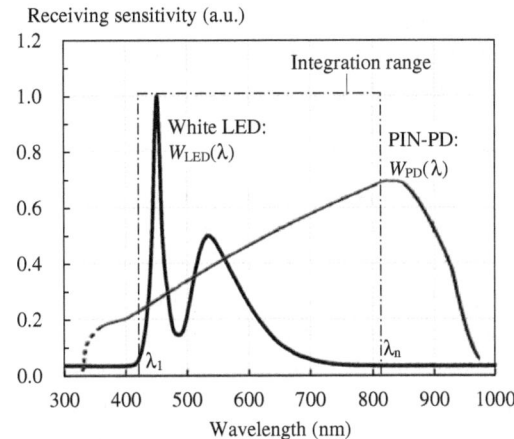

For VLC systems that working in indoor environments, ambient light noises mainly contain intense VL radiation arising from sunlight, skylight, incandescent, fluorescent lamps and so on. The optical intensity spectra of some common VL sources are shown in Fig. 4.8.

Where Sunlight and incandescent lamps are essentially unmodulated sources that can be received at an average power much larger than the desired signal, even when optical filtering is employed.

Fluorescent lamps emit strongly at spectral lines of in 380–700 nm band of interest for VLC systems. Fluorescent-lamp emission is modulated in a near-periodic fashion at the lamp drive frequency, and the detected electrical power spectrum contains discrete components at harmonics of the drive frequency. Traditional fluorescent lamps have been

Fig. 4.8 Light intensity spectra of some VL sources

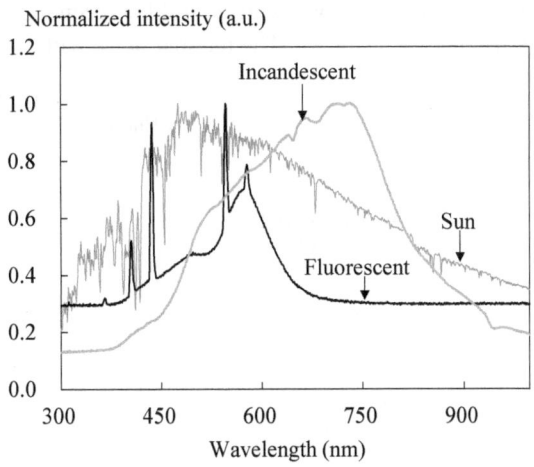

driven at the power-line frequency (50 or 60 Hz), and their electrical spectrum has contained energy at harmonics up to tens of kilohertz [10]. However, recently introduced, high-efficiency *electronic ballasts* drive the lamps at frequencies of tens to hundreds of kilohertz. Their detected electrical spectrum contains energy up to hundreds of kilohertz [10, 11], making electronic-ballast fluorescent lamps are potentially much more detrimental to VLC links than their conventional-ballast counterparts. The system penalty caused by fluorescent-light noise depends strongly on the modulation scheme employed. For a given received fluorescent optical power value, PPM suffers much smaller penalties than OOK. And in principle, subcarrier modulation (see Sect. 5.3) can be made immune to fluorescent-light noise, by choosing the subcarrier frequency high enough [1].

Assuming that the desired signal and ambient light are received with average optical powers P_r and P_n, respectively, they result direct current (DC) photocurrent and induce in the photodetector a *shot noise* current, which is essentially white and Gaussian, so the shot-noise variance (i.e., the first term of (4.4)) is given by

$$\sigma^2_{shot} = 2q(i_r + i_d)RB_n = 2q(P_r + P_n)RB_n, \tag{4.7}$$

where, q is electronic charge and $B_n = I_2R_b$ is noise bandwidth, which $I_2 = 0.562$ is defined as a noise-bandwidth factor, and R_b is the bit rate of the VLC system.

The shot noise of VLC receiver is caused by *light*.

4.3.2 Thermal Noise

The thermal noise (or Johnson noise) that in (4.4) is caused by external electronic circuit of the photodetector. The transimpedance amplifier circuit is popularly used in the receiver of VLC system, as shown in Fig. 4.9 (v_{in} and v_{out} indicate the input and output voltage). This approach can control the receiving sensitivity and bandwidth by employing a feedback resistor R_F. The thermal noise with R_F is given by

$$\sigma^2_{thermal} = \frac{4kTB_n}{R_F}. \tag{4.8}$$

Fig. 4.9 The transimpedance amplifier circuit for a VLC receiver

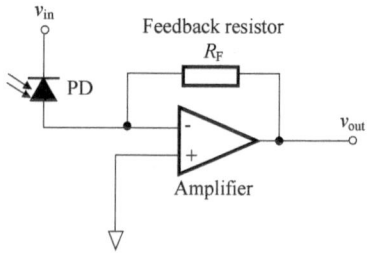

Here, k is Boltzmann's constant ($k = 1.380658 \times 10^{-23}$ J/K) and T(K) is absolute temperature of ambient environment. Because of the thermal noise of the feedback resistor the receiver noise level is higher and hence the sensitivity is degraded. However, sensitivity can be improved by increasing the value of R_F. On the other hand, in order to extend the bandwidth R_F has to be low.

The thermal noise of VLC receiver is caused by *heat*.

4.3.3 Receiver SNR

Both the transmission quality, i.e., the receiver BER, and capacity of wireless communication systems depend on the SNR (dB) at its receiver. The SNR is determined mainly by the receiver and characteristics of spatial channel. In a VLC system, the SNR is the ratio of the received average power P_r (W) with the receiving sensitivity of receiver R (A/W) to the total noise output P_N (A^2); the receiver's sensitivity directly effects the theoretical received power. The SNR is expressed as

$$\text{SNR} = \frac{(RP_r)^2}{P_N}. \tag{4.9}$$

By (4.4), (4.7), and (4.8),

$$P_N = 2q(P_r + P_n)RB_n + \frac{4kTB_n}{R_F}. \tag{4.10}$$

And the BER, the most important value for evaluating digital transmission quality, is given by

$$\text{BER} = Q\left(\sqrt{\text{SNR}}\right), \tag{4.11}$$

where

$$Q(x) = \frac{1}{\sqrt{2\pi}} \int_x^{\infty} e^{-y^2/2} dy (x \geq 0). \tag{4.12}$$

Since BERs strongly depends on the modulation scheme, BERs corresponding to various modulation schemes will be discussed in detail in Sect. 5.2.1.

In addition using the SNR, a theoretical upper limit on the data transmission rate of a communication channel can be given by the Shannon–Hartley theorem (also called Shannon's theorem)

$$C = B \cdot \log_2(1 + \text{SNR}), \tag{4.13}$$

Fig. 4.10 Channel capacity
per unit bandwidth versus SNR

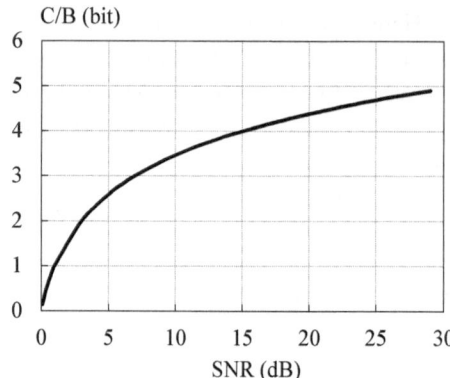

where C is the channel capacity, which represents the maximum rate of reliable data transmission in bits per second (bps) over the channel, B is the bandwidth of the channel in Hz, which refers to the range of frequencies that the channel can accommodate. The theorem provides a formula for the maximum rate at which error-free data can be transmitted over a channel with a specified bandwidth and SNR. Figure 4.10 shows the logarithmic change of channel capacity per unit bandwidth with the SNR. That is, as the SNR increases, the change tends to be slow and the dependence on SNR weakens.

Shannon's theorem essentially states that the channel capacity is directly proportional to the bandwidth and the logarithm of the signal-to-noise ratio. This implies that the higher the bandwidth or the signal-to-noise ratio, the greater the channel capacity and hence the potential data transmission rate. For a wireless communication system, such as a VLC system, if an appropriate modulation method (has good bandwidth and SNR) is used, then data can be transmitted by using maximum communication capacity of this system.

4.3.4 Indoor Multipath Effect

For indoor VLCs, multipath effects caused by reflected light from walls or plural lamp sources is possible. Figure 4.11a shows the reflected-light multipath effect. LED lamps are generally installed on the ceiling and the illumination light has larger diffusion range, i.e., a nondirected-LOS link. In this case, the receiver receives not only direct light from the LED lamp, but also the reflected light from walls, which have propagation delay time. The influence of the direct light depends on performance of VLC system, and the reflected light depends on the reflectance of walls and the length of the reflection path. Despite this, some studies have shown [12, 13] that the influence of the direct light is much larger than reflected light. Therefore, the influence of reflected light can generally be ignored.

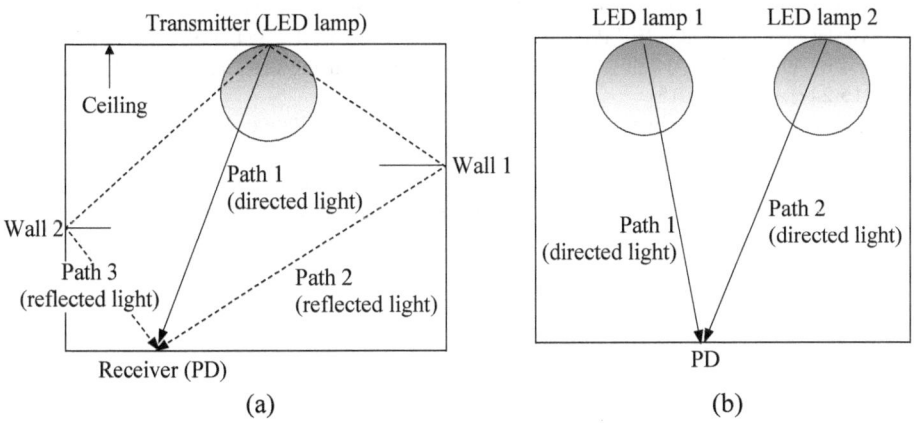

Fig. 4.11 The multipath effect is caused by: **a** reflective light from walls and **b** plural lamp sources

Figure 4.11b shows the plural-sources multipath effect. In this case, the appropriate configuration of plural LED lamps can avoid or mitigate the interference from multipath sources. For example, by leaving enough distance between adjacent lamps, one receiver can only receive signals from one lamp. On the other hand, the inherent robustness of the OFDM (orthogonal frequency division multiplexing, see Sect. 5.3.1) against multipath effects can make it a good choice of modulation scheme in this case [14].

References

1. J.M. Kahn, J.R. Barry: Wireless infrared communications, Proc. of the IEEE **85**(2), 256–298 (1997)
2. T. Takano: Lightwave wireless communications and infrared technologies, J. Jpn. Soc. Infrared Science & Technology **13**(2), 6–11 (2004)
3. V. Manea, R. Dragomir, S. Puscoc: OOK and PPM modulations effects on bit error rate in terrestrial laser transmissions, Telecomunicatii **2011**(2), 55–61 (2011)
4. J. Sakai: Chapter 14. In: Optical Communication Engineering, (Kyoritsu Shuppan, Tokyo 2000) pp.133–144
5. A. Selvarajan, S. Kar, T. Srinivas: Chapter 8. In: Optical Fiber Communications, ed. by G. Keiser (McGraw-Hill, New York 2002) pp.140–149
6. https://www.rp-photonics.com/optical_heterodyne_detection.html
7. Y. Sugimori, W. Sakamoto: Chapter 2 and 3. In: Marine Environmental Optics (TokaiUnivPress, Hiratsuka-shi 1985) pp. 41–128
8. T. Aruga: Chapter 3. In: Spatial Transmission Optics (Suiyosha, Tokyo 2000) pp. 61–94
9. D. L. Rogers: Integrated optical receivers using MSM detectors, IEEE J. Lightwave Tech. **9**(12), 1635–1638 (1991)
10. A. J. C. Moreira, R. T. Valadas, A. M. de Olveira Duarte: Optical interference produced by artificial light, Wireless Networks **3**, 131–140 (1997)

11. R. Narasimhan, M. D. Audeh, J. M. Kahn: Effect of electronic-ballast fluorescent lighting on wireless infrared links, IEE Proc. -Optoelectronics **143**(6), 347–354 (1996)
12. X. Lin: Visible-light wireless communications technique using LED lighting, IEICE Tech. Rep. **115**(247), 63–68 (2015)
13. T. Komine, M. Nakagawa: Fundamental analysis for visible-light communication system using LED lights, IEEE Trans. Consum. Electron. **50**(1), 100–107 (2004)
14. Y. Tanaka, T. Komine, S. Haruyama, M. Nakagawa: Indoor visible communication utilizing plural white LEDs as lighting. In: 12th IEEE Int. Symp. PIMRC 2001 (2001) pp. F81–F85

VLC Modulation Techniques

5

According to different usage purposes, VLC systems can choose either BIM (baseband intensity modulation) or SCM (subcarrier modulation) to modulate the carrier signal. This chapter mainly introduces various modulation schemes commonly used in VLC systems in these two modulation techniques, and the different effect of various modulation schemes of BIM on BER (bit error rate), bandwidth requirement, and flicker generated by optical modulation is also discussed. Additionally, the VLC-specific CSK (color-shift keying) scheme is also introduced.

Modulation techniques for VLC systems include baseband intensity modulation (BIM) and subcarrier modulation (SCM), as well as some generalizations of these techniques. The BIM is typically used in scenarios where simplicity and cost-effectiveness outweigh the need for extremely high data rates and noise resistance, such as short-distance communication or in situations where the available bandwidth is sufficient for the application's requirements. SCM can be more complex to implement than single-carrier BIM schemes. However, SCM allows for efficient utilization of the available bandwidth by dividing it into multiple subchannels, and it provides resistance to frequency-selective fading and can mitigate the effects of interference. So is well-suited for high-data-rate transmission in communication systems.

X. Lin, *Visible Light Communications*, Synthesis Lectures on Communications,
https://doi.org/10.1007/978-3-031-64475-7_5

5.1   Baseband Intensity Modulation (BIM)

BIM is most viable modulation technique used in VLC systems to transmit digital or analog information by varying the intensity or power level of a continuous-wave carrier signal. BIM consists of a carrier signal source (CSS), an information signal source (ISS), and a modulator, as shown in Fig. 5.1.

CSS is typically a high-frequency continuous-wave signal. In VLC systems, CSS general is an LED or an LD light source. ISS, i.e., baseband signal is a signal (analog or digital) that carries the data to be transmitted, it used to control and vary the power or intensity of the carrier signal directly. The modulator is the key component responsible for varying the intensity of the carrier signal according to the information signal. The intensity amplitude of the carrier signal is varied directly in proportion to the instantaneous amplitude of baseband ISS. The modulated carrier signal is then transmitted through the channel, which is a free space in the case of VLC systems. As the modulated carrier signal travels through the channel, variations in its intensity represent the encoded information.

On–off keying (OOK) using non-return-to-zero (NRZ) pulse (OOK-NRZ), pulse-position modulation (PPM) with L levels (LPPM), inverted-LPPM (I-LPPM), and pulse-width modulation (PWM) are the most desired and common baseband signal schemes of BIM for VLC systems [1]. Their coding methods as shown in Fig. 5.2, from top to bottom are OOK-NRZ, PWM, 4PPM, and I-4PPM.

Besides that, Optical duobinary modulation (DBM) is also a BIM technique used in optical communication systems to encode digital information onto an optical carrier signal for transmission over optical fiber networks or spatial channel. It is a form of modulation that helps improve the performance of high-speed optical communication systems by reducing the effects of chromatic dispersion and other impairments.

Here are their principles and applications, respectively.

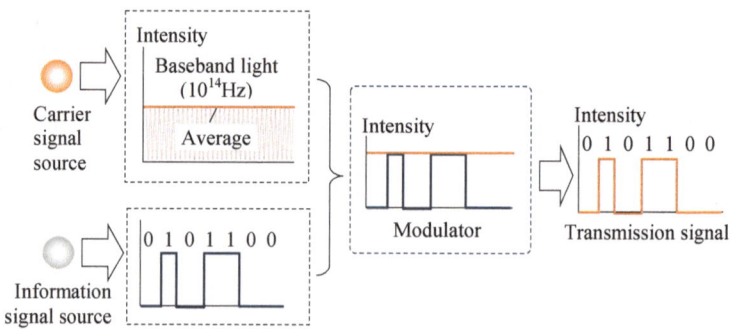

Fig. 5.1 Principle of BIM technique

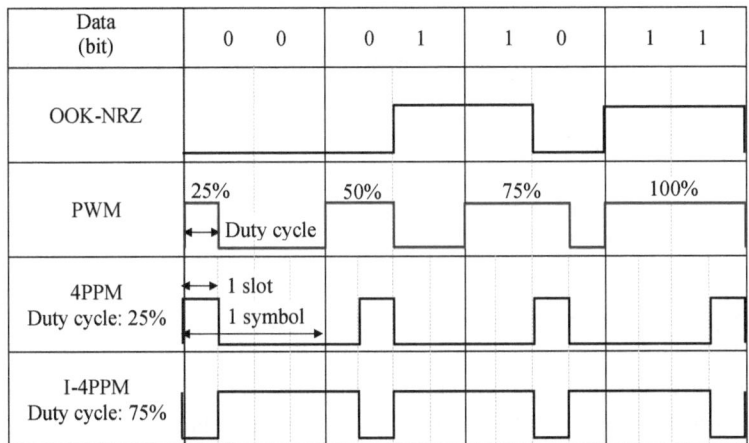

Fig. 5.2 Coding methods for OOK-NRZ, PWM, 4PPM, and I-4PPM

5.1.1 On–Off Keying with Non-Return-to-Zero (OOK-NRZ)

OOK-NRZ is the simplest scheme among all BIM techniques suitable for VLC systems, it is essentially a combination of two key concepts: NRZ and OOK.

In NRZ modulation, each binary bit is represented by a signal level that remains constant for the duration of the bit. A high-level signal (typically a high voltage or light intensity) represents binary 1 state, and a low-level signal (usually zero voltage or no light) represents binary 0 state. There is no return to a zero-signal level during the bit period, hence the name "Non-Return-to-Zero (NRZ)".

In OOK modulation, the presence or absence of a carrier signal represents the binary data. When a binary 1 is to be transmitted, the carrier signal is turned on (typically at full strength), and when a binary 0 is to be transmitted, the carrier signal is turned off (completely suppressed).

OOK-NRZ combines these two concepts by modulating the presence and absence of the carrier signal according to the NRZ-encoded binary data. When a binary 1 is encountered in the NRZ-encoded data, the carrier signal is transmitted at full strength. When a binary 0 is encountered, the carrier signal is completely turned off. The resulting modulated signal consists of bursts of carrier signal (high level) for binary 1 s and periods of no signal (low level) for binary 0 s, as shown in the top row of Fig. 5.2.

OOK-NRZ is straightforward to implement and requires relatively simple hardware. Also, it does not require additional bandwidth for synchronization purposes (e.g., clock signals) since it directly represents the binary data, hence it has good bandwidth efficiency. However, OOK-NRZ is susceptible to noise and interference, which can affect the accuracy of data transmission.

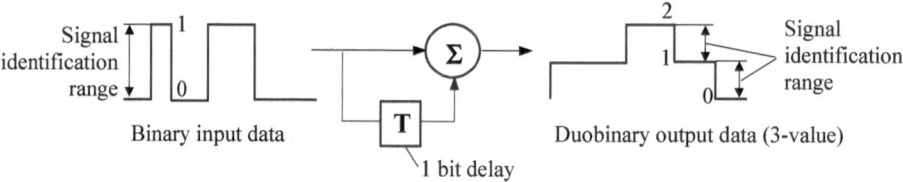

Fig. 5.3 The generation of the duobinary code

5.1.2 Optical Duobinary Modulation (ODBM)

The generation of the duobinary code is performed using an adder that is shown in Fig. 5.3 with a 1-bit delay on input data (binary: 1/0). The output data will be 3-level (3-value) data [2]. This multilevel encoding can improve the bandwidth efficiency of the signal since it transmits multiple bits per symbol.

In an ODBM signal, level 0 and 2 produce 100% transmission with opposite optical phases, and level 1 produces 0% transmission [3–5]. ODBM has a narrow spectral width and large dispersion tolerance due to multilevel encoding. Hence, it is a better scheme to transmit high-speed optical signals over bandwidth-limited channel. However, also due to the multilevel encoding, the identification range of the signal becomes narrower, that result the bit error rate (BER) is increased. Also, implementing duobinary modulation can be more complex than simpler modulation schemes like OOK because it involves pulse shaping and more sophisticated signal processing at both the transmitter and receiver.

In general, ODBM is a better choice for uncompensated single-mode fiber communications, since it is more resilient to chromatic dispersion. ODBM is often used in conjunction with other advanced modulation techniques and error correction coding schemes to further enhance the performance of optical communication systems. The ODBM can also be suggested and used for WDM-based VLC systems for which long distance or/and high speed are required.

5.1.3 Pulse-Position Modulation (PPM)

PPM is a digital modulation scheme used in data transmission, particularly in applications where timing accuracy and resistance to noise are critical. In PPM, data is represented in the form of discrete symbols. Each symbol corresponds to a specific pattern of pulses within a given time interval. PPM modulates the position of pulses within a fixed time interval, often referred to as the symbol period (1 symbol). The position of the pulse within this period indicates the value of the symbol. Each symbol in PPM is associated with a particular time slot or position within the symbol period. The presence or absence of a pulse at a specific position within the symbol period indicates the binary value of that symbol.

PPM is more robust against amplitude variations and noise compared to some other modulation schemes. Since the information is encoded in the pulse position, changes in amplitude due to noise may not affect decoding accuracy as much. Also, because PPM requires precise timing synchronization between the transmitter and receiver, this makes it suitable for applications where timing accuracy is crucial.

PPM with L levels ($LPPM$) is also widely used in illumination-light-based VLC system, where L is number of *slots* in 1 *symbol*. Third row of Fig. 5.2 shows a case of L = 4. In order to reduce the amount of light loss due to light intensity modulation, the inverted-$LPPM$ waveform can be used. I-$LPPM$ is logical inverted $LPPM$. The bottom of Fig. 5.2 shows the case of $L = 4$. In this case, each symbol has three-quarters of the amount of all light and 4PPM only has one quarter. The consideration to ensure as much of the light as possible is invariant (i.e., light illumination is invariant) is also important for a lighting-based VLC system. Hence, I-4PPM is approved by JEITA (Japan electronics and information technology industries association) as a standardized BIM technique for VL identification (ID) systems [6] and VL beacon systems [7].

5.1.4   Pulse-Width Modulation (PWM)

PWM is also an intensity modulation scheme, which modulates light intensity by changing the duty cycle of the pulse wave, as shown in Fig. 5.4, the stronger the voltage signal, the larger the duty cycle of the PWM signal. Duty cycle is measured in percentage. The percentage duty cycle specifically describes the percentage of time a digital signal is on over an interval or period of time. If a digital signal spends half of the time on and the other half off, this digital signal has a duty cycle of 50% and resembles an ideal square wave. If the percentage is higher than 50%, the digital signal spends more time in the high state than the low state and vice versa if the duty cycle is less than 50%, as shown in second row of Fig. 5.2.

The PWM method can be used to control brightness (i.e., dimming) of LEDs by adjusting the duty cycle. When an LED lamp is used as sending source in a VLC system, the illumination dimming maybe required. In this case, how to incorporate dimming while not corrupting the communication link is an important subject. *Ntogari* et al. proposed a method of combining PWM and discrete multitone (DMT), which can ensure dimming

Fig. 5.4   Method of PWM

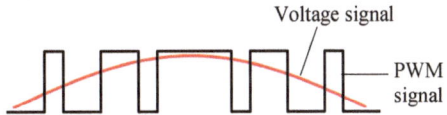

does not influence the data transmission [8]. Also, PWM doesn't require complex circuitry and is easy to implement.

5.2 Transmission Quality in BIM

The three most important criteria for evaluating BIM techniques are [9]:

- The average received optical power required to achieve a desired BER
- The bandwidth requirement of the receiver
- Flicker efficiency (i.e., reduction of flicker caused by optical modulation)

5.2.1 Receiver BER and Bandwidth Requirement

The receiver BER, the most important value for evaluating a digital transmission quality, the probability that an error may occur in a bit in the pulse train, i.e., a "1" bit turns into a "0" bit or vice versa. The BER is determined by the modulation scheme and the receiver SNR. When the light modulation by using the OOK-NRZ and ODBM scheme, the BERs are given by (5.1) and (5.2), respectively [9, 10]

$$\text{BER}_{\text{OOK-NRZ}} = Q\left(\sqrt{\text{SNR}}\right) \tag{5.1}$$

and

$$\text{BER}_{\text{duo-binary}} = \frac{16}{21}Q\left(\frac{\sqrt{\text{SNR}}}{2}\right) \tag{5.2}$$

And when by using the $LPPM$ and I-$LPPM$ scheme, the BERs are given by (5.3) and (5.4), respectively [11]

$$\text{BER}_{LPPM} = \frac{L}{2}Q\left(\sqrt{\frac{L\log_2 L}{2}}\sqrt{\text{SNR}}\right) \tag{5.3}$$

and

$$\text{BER}_{\text{I}-LPPM} = \frac{L}{2}Q\left(\sqrt{\frac{L\log_2 L}{2(L-1)}}\sqrt{\text{SNR}}\right), \tag{5.4}$$

where

$$Q(x) = \frac{1}{\sqrt{2\pi}}\int_x^\infty e^{-y^2/2}dy(x \geq 0) \tag{5.5}$$

Table 5.1 Comparisons of modulation characteristics for $LPPM$, OOK-NRZ, and DBM. R_b is bit rate, P_a is average received power

	Bandwidth requirement	Power requirement	Signal identification ability
$LPPM$	$(R_b/2)(L/\log_2 L)$ R_b (when $L = 4$)	$P_a/\sqrt{(L/2)(\log_2 L)}$ $P_a/2$ (when $L = 4$)	High
OOK-NRZ	$R_b/2$	P_a	Identification range
ODBM	$R_b/3$	$2\,Pa$	Low Identification range

When $L = 2$ in (5.3), $\mathrm{BER}_{OOK\text{-}NRZ} = \mathrm{BER}_{LPPM}$, that is, 2PPM is equivalent to OOK-NRZ. This also can be derived from Table 5.1. Power requirement can also readily be derived from the BER expressions. From Table 5.1, $LPPM$ requires a factor of $\sqrt{(L/2)(\log_2 L)}$ less power than OOK-NRZ to obtain a particular BER performance.

Table 5.1 compares modulation characteristics for $LPPM$, OOK-NRZ, and ODBM in terms of bandwidth, power, and signal identification ability [12]. From Table 5.1, for a given bit rate R_b, $LPPM$ requires more than OOK-NRZ and DBM by a factor $L/\log_2 L$, e.g., 4PPM requires two times more bandwidth than OOK-NRZ (also see Fig. 5.5). Therefore, the OOK-NRZ scheme has a higher data transmission rate, and it is almost bit rate-independent for data rates in excess of 100Mbps, and $LPPM$ is a preferred modulation scheme for devices where low-power consumption is required [11], such as portable and underwater optical transmitters due to its high-power efficiency.

$LPPM$ yields an average power requirement that decreases steadily with increasing L. When L is larger than 2, the $LPPM$ exhibits a higher efficiency than OOK-NRZ and ODBM does. However, an increase in L (i.e., in power efficiency) causes an increase in the bandwidth requirement.

Since the multilevel encoding, the signal identification range of ODBM is narrower than $LPPM$ and OOK-NRZ, which are binary-level encoding (also see Fig. 5.3), it makes the BER of ODBM is lager.

Figure 5.6 shows the simulation results of BER versus SNR by using (5.2)–(5.4) corresponding to various modulation schemes. Figure 5.6a is ODBM, OOK-NRZ, 4PPM, and I-4PPM, and Fig. 5.6b is $LPPM$s when $L = 2, 4, 6, 8$.

From Fig. 5.6a, ODBM has larger BER, which same as the discussions in Table 5.1. From Fig. 5.6b, for $LPPM$s with different L, the SNR requirement to reach a certain

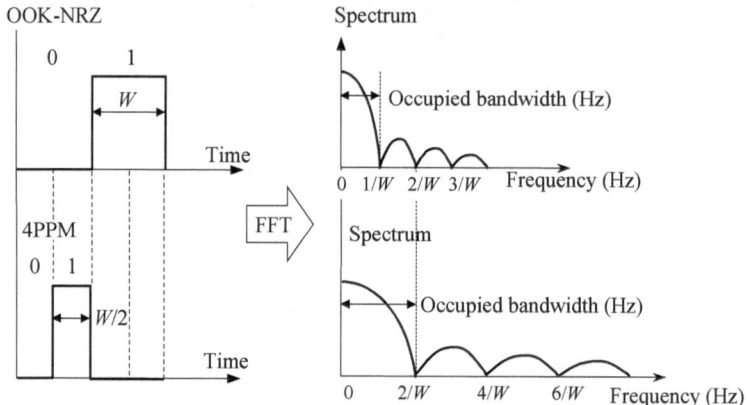

Fig. 5.5 Occupied bandwidth for OOK-NRZ and LPPM. FFT: fast fourier transform

BER (e.g., 10^{-8}, for general VLC systems, BER $= 10^{-8}$ is a typical required value) is decreased with increasing L, and has higher data transmission efficiency than the OOK-NRZ. However, note that an increase in the L causes a decrease in the communication speed.

5.2.2   Flicker in VLC

Flicker is a physiological optics phenomenon where the light source repeats light and dark phases moderately enough to be perceived by the human eyes, it will cause varying-degrees sense of discomfort, so the flicker of the light source should be minimized. In VLC systems, flicker is caused by intensity modulation to light source, and it is related to modulation scheme, speed, depth, and light-source color [13].

Figure 5.7 shows the flicker versus modulation speed and depth. High speed makes the light intensity nearly constant, as shown in Fig. 5.7a, and shallow modulation depth makes the light intensity even, as shown in Fig. 5.7b. So in both cases, the flicker is relatively weak. The luminous color of the light source is also one of the factors that affects flicker. Generally, white light has a stronger flicker effect than monochromatic light [13].

Fig. 5.6 BER versus SNR
corresponding to various
modulation schemes:
a ODBM, OOK-NRZ,
I-4PPM, and 4PPM and
b *L*PPMs when *L* = 2, 4, 8, 16

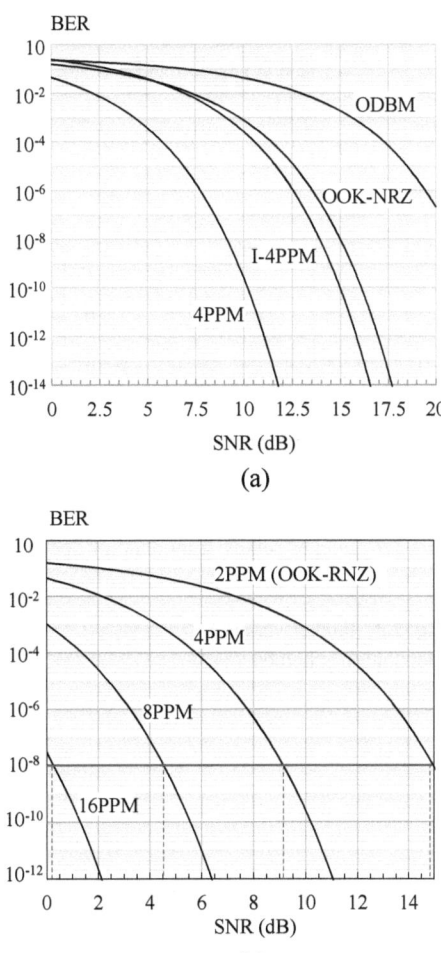

The flicker versus modulation schemes is shown in Fig. 5.8. In *L*PPM (and I-*L*PPM), light power does not change within every symbol period, so it has a weaker flicker than others which the light power changes within every symbol period. Low flicker makes *L*PPM and I-*L*PPM suitable to be employed in indoor lighting-based VLC systems. The DBM has also a small flicker because its modulation depth is shallow, although there is a change in light power within its symbol period. Flicker mitigation is one of main challenges for VLC. The IEEE Standard 802.15.7 [14] have been given some solutions for flicker mitigation, and they will be described in Sect. 7.2.3.

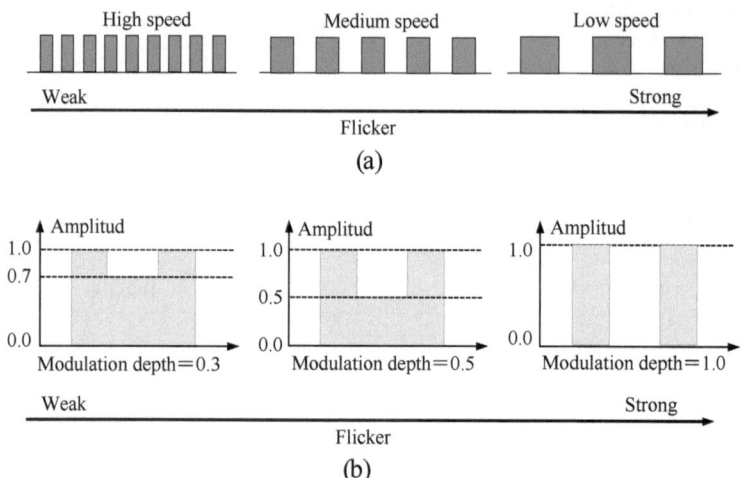

Fig. 5.7 Flicker versus: **a** modulation speed and **b** modulation depth

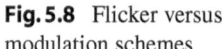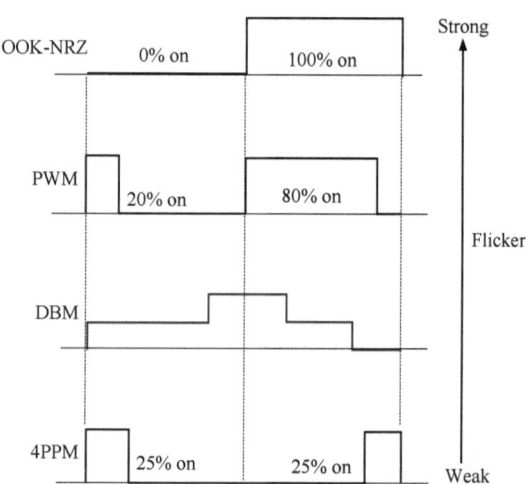

Fig. 5.8 Flicker versus modulation schemes

5.3   Subcarrier Modulation (SCM)

SCM is a modulation technique commonly used in various communication systems which include VLC. The key components in SCM are a subcarrier (SC) generator and a SC modulator, as shown in Fig. 5.9.

The generation of subcarriers is typically done using frequency division techniques or by using specialized subcarrier generation circuits. generated SC is a radio wave signal

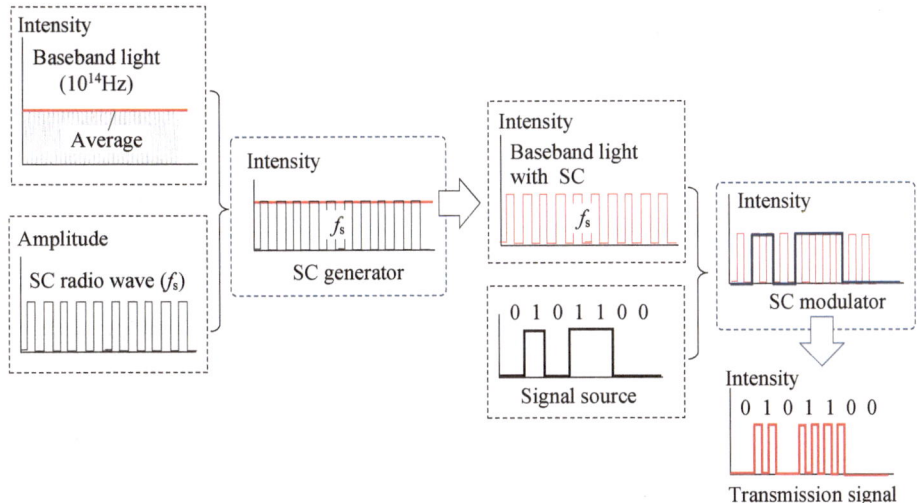

Fig. 5.9 Principle of SCM technique

with subcarrier frequency (SCF) f_s. The SCF can be selected according to different communication purposes. For example, the SCF for a VL-ID transmission system is usually 15–40 kHz [7], for multimedia communication such as portraits and sounds is usually above 1 MHz, and can also by selecting the corresponding SCF to avoid the ambient noise.

Since the SCF is a radio wave frequency, the SC modulator can use any suitable modulation scheme for radio wireless systems include amplitude, frequency, and phase modulation (AM, FM, and PM), as well as some generalizations of these techniques or more advanced modulation schemes. Since digital communication only switches between 0 and 1 binary values, the corresponding AM, FM, and PM are called ASK, FSK, and PSK (amplitude, frequency, and phase shift keying) respectively, as shown in Fig. 5.10 [15].

SCM can be more complex to implement than BIM schemes, however since it allows for efficient utilization of the available bandwidth by dividing it into multiple subchannels, also it provides resistance to frequency-selective fading and can mitigate the effects of interference, so SCM widely used in various modern, well-robustness VLC systems. Especially orthogonal frequency division multiplexing (OFDM) scheme is widely used in various high-data-rate VLC systems include VLC-based Li-Fi (Light Fidelity) system, due to its ability to efficiently transmit data over wide bandwidths and its robustness in dealing with channel impairments.

Fig. 5.10 Common binary modulation schemes for radio wireless systems

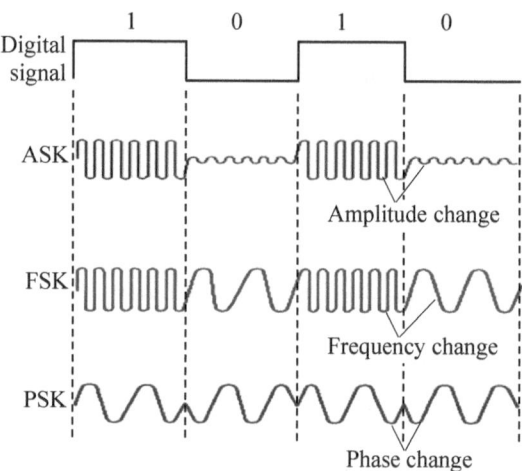

5.3.1 Orthogonal Frequency Division Multiplexing (OFDM)

OFDM is a digital multicarrier parallel modulation scheme, which extends the concept of single subcarrier modulation by using multiple subcarriers within the same single channel to realize large-capacity data transmission. It works on the principal of FDM (frequency division multiplexing). In FDM, data are transmitted in parallel on a number of different frequencies, and each frequency channel is separated from the others by a frequency *guard band* in which no information is transmitter to reduce intersymbol interference between adjacent channels. as shown in Fig. 5.11a.

OFDM on the other hand uses an orthogonal carrier which allows overlapping of band without interfering since orthogonal vectors don't overlap and since signals are vectors it uses orthogonal carrier. The orthogonality is achieved by first dividing serial data into parallel symbol increasing effective symbol period then multiplying it by IFFT (inverse fast Fourier transform) matrix. Since Guard band is avoided and overlapping of frequency is allowed as shown in Fig. 5.11b, a large amount of bandwidth is saved and the bandwidth is doubled nearly. Currently, OFDM is used in most broadband wired and wireless communication systems. However, in these radio frequency transmission systems, the high peak-to-average power ratio (PAPR) in OFDM is usually considered a disadvantage due to nonlinearities of the power amplifier.

Recently, a number of studies have shown that OFDM is also a promising technology for OWCs [16, 17]. *Afgani* et al. demonstrated theoretically that the high PAPR in OFDM can be exploited constructively in an LED-based VLC system to intensity modulate the light from LED to realize high-speed VL wireless links [18]. However, the nonlinear behavior of LEDs in the transmitter is a limiting factor of system performance. Optical modulation signals with large PAPR suffer uneven distortion.

Fig. 5.11 OFDM compared to FDM for multiband: **a** conventional FDM and **b** OFDM. SCB: subcarrier bandwidth

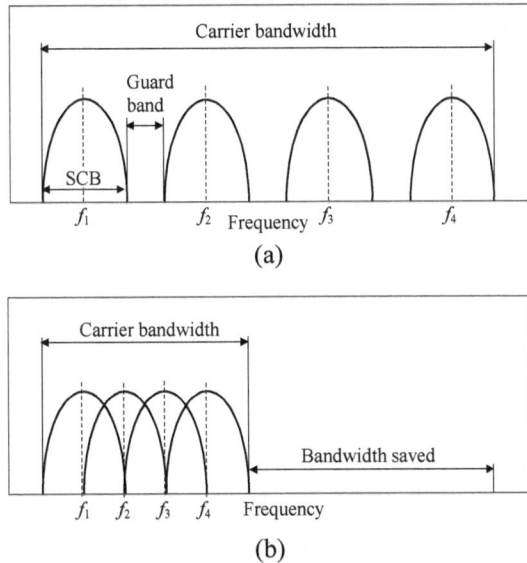

On the other hand, direct biased optical OFDM (DCO-OFDM) and asymmetrically clipped optical OFDM (ACO-OFDM) [19] are two effective intensity-modulation schemes for IM/DD systems. To realize OFDM by IM/DD in OWC systems, the transmitted signal must be non-negative and real. The DCO-OFDM Scheme generates a non-negative polarity signal by adding a DC bias to the baseband OFDM signal, and transmits this signal, as shown in Fig. 5.12a. The ACO-OFDM scheme is a method that can transmit only non-negative polarity signals without adding DC bias by generating signals using only subcarriers with frequencies that are odd multiples of the baseband OFDM frequency, as shown in Fig. 5.12b. To achieve good performance of OFDM scheme in optical systems, OFDM technique must be adapted in various ways.

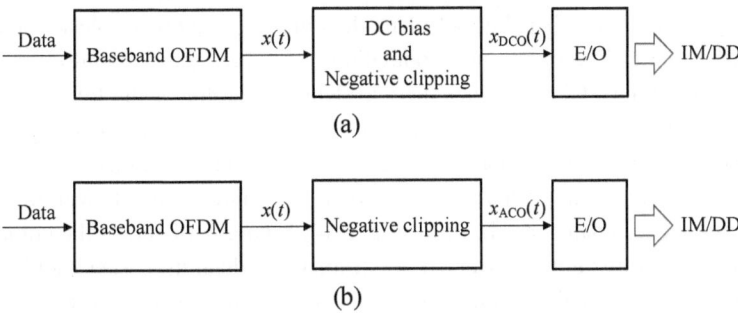

Fig. 5.12 OFDM for IM/DD systems: **a** DCO-OFDM and **b** ACO-OFDM scheme

Fig. 5.13 Method of DMT

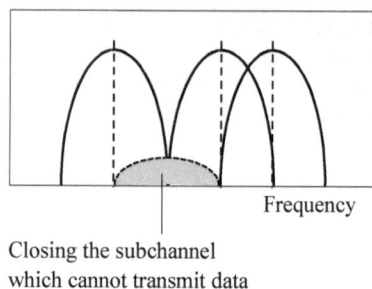

Frequency

Closing the subchannel
which cannot transmit data

5.3.2 Discrete Multitone (DMT)

DMT is a variant of OFDM, its basic idea is also to split the available bandwidth into a large number of subchannels. The study of *Qian* et al. shows that DMT can be well suited to nonlinear LED-based VLC system [20]. DMT has good system performance since it is an orthogonal linear transformations scheme, it can spread the nonlinear effects evenly to each data symbol. Also, DMT can allocate data so that the throughput of every single subchannel is maximized. If some subchannel cannot carry any data, it can be turned off and the use of available bandwidth is optimized, as shown in Fig. 5.13.

DMT systems continuously monitor the channel conditions, such as noise and interference levels, and adapt the modulation and coding parameters for each subchannel dynamically. This adaptive bit loading ensures that the available bandwidth is used efficiently and that the system can provide the highest possible data rates without excessive errors. *Vucic* et al. have realized a high-speed LED-based VLC of 513Mbps using DMT modulation [21].

5.4 Color-Shift Keying (CSK)

Color-shift keying (CSK) is a new VLC modulation scheme approved as part of IEEE 802.15.7−VLC in 2011 [14]. Instead of modulating the amplitude, phase, or frequency of the carrier signal like traditional modulation techniques (e.g., amplitude-shift keying or phase-shift keying), CSK changes the wavelength or color of the transmitted light to encode digital information. It is a technique that takes advantage of the properties of optical filters and the spectral characteristics of light to transmit information.

The CSK signal is generated by three-color LV sources, and transmitting data by using the color coordinates, as shown in Fig. 5.14. Figure 5.14 shows the color codes of CSK on a X–Y chromaticity diagram. The chromaticity diagram is specified by CIE1931 (Commission International de l'Eclairage). The color of light is represented by the plane coordinates of X, Y. The light wavelength is written around the curve (i.e., spectral locus) of the chromaticity diagram. Each monochromatic is represented by these wavelengths,

Fig. 5.14 Color codes of CSK on X–Y chromaticity diagram

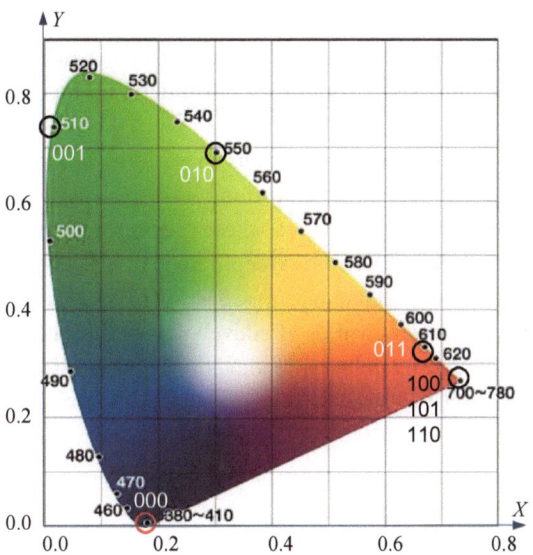

with wavelengths shown in nanometers. The horizontal axis X and the vertical axis Y are values of chromaticity. The circle mark on the spectral locus indicates a code of data, for example, the monochromatic wavelength of 510 nm corresponds to code 001. In CSK scheme, the communication rate is determined by the number of constellations on the chromaticity diagram, and the speed increases with increasing constellation.

Here are the key aspects of CSK modulation:

- Multi-level encoding: CSK can be implemented using various levels of wavelength shifts. Binary CSK (BCSK) uses two wavelengths to represent binary data, while multilevel CSK (MCSK) can use more than two wavelengths for higher data rates. In MCSK, each color represents a group of bits, allowing for more efficient use of the optical spectrum.
- Spectral efficiency: CSK can be more spectrally efficient than some other modulation techniques, as it can transmit multiple bits simultaneously using different wavelengths. This spectral efficiency is especially valuable in optical communication systems where bandwidth is limited.
- Dispersion tolerance: CSK can provide better tolerance to chromatic dispersion compared to amplitude-based modulation schemes like ASK.
- Coding and error correction: Like other modulation techniques, CSK can benefit from error correction coding to improve the reliability of data transmission, especially in noisy or challenging VL environments.

– Complexity: Implementing CSK requires wavelength-selective components like filters or gratings to separate and detect different colors at the receiver. This can add complexity to both the transmitter and the receiver.

CSK has found applications in various optical communication systems, including free-space optical communication and wavelength-division multiplexing (WDM) systems. It can be used in scenarios where spectral efficiency and tolerance to dispersion are critical, and it is suitable for video-based VLC systems that have image-processing or color-control requirement [22]. CSK is a special and interesting modulation scheme limited to VL carriers, which effectively leveraging the color characteristics of VL. As one of the important modulation methods adopted in VLC international standardization, it will also be described from different perspectives in Sect. 7.2.

References

1. J.M. Kahn, J.R. Barry: Wireless infrared communications, Proc. of the IEEE **85**(2), 256–298 (1997)
2. *ntoshio.la.coocan.jp/rakuen/10basic/opt03/index.htm*
3. N. Fujimoto: The fastest modulation method for visible light communications, News Release of Kindai University, 15th October (2012)
4. K.-P. Ho: Optical duobinary modulation, Recent Pat. Eng. **4**(2), 80–85 (2010)
5. R. Kaur, S. Dewra: Duobinary modulation format for optical system--a review, Int. J. Adv. Res. Electr. Electron. Instrum. Eng. **3**(8), 11039–11046 (2014)
6. JEITA: *CP-1222: Visible Light ID System* (JEITA, Tokyo 2007)
7. JEITA: *CP-1223: Visible Light Beacon System* (JEITA, Tokyo 2013)
8. G. Ntogari, T. Kamalakis, J. Walewski, T. Sphicopoulos: Combining illumination dimming based on pulse-width modulation with visible-light communications discrete multitone, J. Opt. Commun. Netw. **3**(1), 56–65 (2011)
9. X. Lin: Visible-light wireless communications technique using LED lighting, IEICE Tech. Rep. **115**(247), 63–68 (2015)
10. H. Sugiyama, S. Haruyama, M. Nakagawa: Analysis and experiment of communication distance in visible light communication system, IEICE Tech. Rep.**106**(598), 25–30 (2007)
11. M.D. Audeh, J.M. Kahn, J.R. Barry: Performance of pulse-position modulation on measured non-directed indoor infrared channels, IEEE Trans. Commun. **44**(6), 654–659 (1996)
12. V. Manea, R. Dragomir, S. Puscoc: OOK and PPM modulations effects on bit error rate in terrestrial laser transmissions, Telecomunicatii **2011**(2), 55–61 (2011)
13. T. Saito, S. Haruyama, M. Nakagawa: A study for flicker on visible light communication, IEICE Tech. Rep. **106**(450), 31–35 (2007)
14. IEEE: Standard 802.15.7–2011: *Short-Range Wireless Optical Communication Using Visible Light* (IEEE, Piscataway 2011)
15. J. G. Proakis, *Digital Communications*, 3rd ed. New York: McGraw-Hill (1995)
16. J. Armstrong: OFDM for optical communications, J. Lightwave Technol.**27**(3), 189–204 (2009)
17. H. Elgala, R. Mesleh, H. Haas: Indoor broadcasting via white LEDs and OFDM, IEEE Trans. Consum. Electron. **55**(3), 1127–1134 (2009)

18. M.Z. Afgani, H. Haas, H. Elgala, D. Knipp: Visible light communication using OFDM. In: *2nd IEEE Intl. Conf. on TRIDENTCOM 2006* (2006), https://doi.org/10.1109/TRIDNT.2006.1649137

19. K. Ohuchi: Fundamental techniques for optical wireless OFDM system, IEICE Fundamentals Rev. **13**(1), 38–46 (2019)

20. H. Qian, S. Cai, S. Yao, T. Zhou, Y. Yang, X. Wang: On the benefit of DMT modulation in nonlinear VLC systems, Opt. Express **23**(3), 2618–2632 (2015)

21. J.Vucic, C. Kottke, S. Nerreter, K-D. Langer, J. W. Walewski: 513Mbit/s Visible light communications link based on DMT-modulation of a white LED, J. Lightwave Technol. **28**(24), 3512–3518 (2010)

22. H. Mizuno, S. Choi, A. Yokoi: Performance of CSK communication systems with displays and cameras, IEICE Tech. Rep. **114**(160), 63–68 (2014)

Optical Multiple-Access Techniques

6

Multiple-access technique is an effective means for achieving high-speed and high-capacity communications. The multiplexing methods suitable for VLC can be classified into two categories based on their physical nature: optical and electrical. This chapter introduces three major optical multiplexing technologies: WDM (wavelength division multiplexing), SDM (space division multiplexing), and PDM (polarization division multiplexing). For TDM (time division multiplexing) and CDM (code division multiplexing) of electrical nature, the methods for how to process the optical signal for realization of optical multiplexing are given. Finally, the OMLM (optical multilevel modulation) scheme of optical multiplexing by using modulation technique is also introduced.

Multiple-access techniques (MATs) are commonly employed in various communication systems, including wireless networks, satellite communication, and Ethernet-based wired networks to allow multiple users or data streams to share a common communication channel efficiently. In this chapter, the optical multiple-access techniques (OMATs) which involves the use of light signals to transmit information are discussed. Efficiently control multiple channels or data streams in optical wireless communication (OWC) systems including VLC is crucial for maximizing the capacity and performance of the communication system. Figure 6.1 shows the concept of OMAT in OWC systems. The essence of MATs is to transmit multiple optical channels or data streams in 1 optical link to achieve high-speed and high-capacity data transmission. MUX (multiplexer) and DEMUX (demultiplexer) are used for multiplexing and demultiplexing respectively.

The nature of OMATs can be classified into two categories: optical and electrical. OMATs with optical nature may permit different channel, i.e., data streams to transmit simultaneously within the same space, without requiring a loss of per-channel capacity (at least in principle), such as wavelength division multiplexing (WDM), space division multiplexing (SDM), and polarization division multiplexing (PDM). For OMTAs with electrical nature, although they can also enable reliable transmission when different channels share

© The Author(s), under exclusive license to Springer Nature Switzerland AG 2025
X. Lin, *Visible Light Communications*, Synthesis Lectures on Communications,
https://doi.org/10.1007/978-3-031-64475-7_6

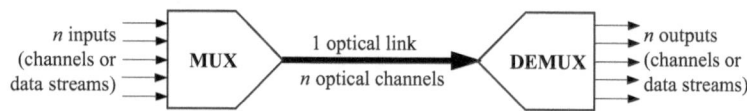

Fig. 6.1 Conceptual diagram of OMAT

Table 6.1 Techniques for multiplexing transmissions (bold denotes an advantage)

Technique	Nature	Necessary loss of per-channel capacity	Optical average power efficiency	Permits simultaneous transmission
WDM	Optical	**No**	**High**	**Yes**
SDM	Optical	**No**	**High**	**Yes**
PDM	Optical	**No**	**High**	**Yes**
TDM	Electrical	Yes	**High**	No
CDM	Electrical	Yes	Moderate	**Yes**
Subcarrier FDM	Electrical	Yes	Low	**Yes**
MLM	Modulation scheme	**No**	Susceptible to noise and interference	**Yes**

the same optical space, but they necessarily entail a loss of per-channel capacity, such as time division multiplexing (TDM), code division multiplexing (CDM), and subcarrier frequency-division multiplexing (FDM). Multilevel modulation (MLM) is a modulation technique using multiple levels for encoding information, such as the ODBM scheme that is discussed in Sect. 5.1.2. These characteristics of techniques are summarized in Table 6.1 [1].

It is worth noting that from Table 6.1, the power efficiency achieved by subcarrier FDM is poor and worsens as the number of subcarriers increases. In subcarrier FDM, different data streams or channels can transmit simultaneously at different subcarrier frequencies, each data stream or channel is assigned a specific frequency band, and they transmit and receive signals within that allocated band. A detailed discussion of subcarrier FDM can be found in Sect. 5.3.1, it will not be repeated in this chapter.

Each OMAT has its advantages and disadvantages, and the choice of a specific technique often depends on factors such as the nature of the application, system requirements, and the desired data rates. In practice, for a VLC system, a combination of these techniques may be used to optimize efficiency and accommodate different types of users and services. Here are more detailed discussions related to various OMATs in Table 6.1.

6.1 Wavelength Division Multiplexing (WDM)

WDM is a common optical multiple-access technique used to increase the data capacity of optical communication systems by simultaneously transmitting multiple signals or channels at different wavelengths (colors) of light. Each wavelength is treated as an independent channel, allowing multiple channels for the parallel transmission of data using a monochromatic light source without interfering with each other, as shown in Fig. 6.2. Here are the key aspects of WDM:

– Wavelengths as channels: in WDM, each signal or communication channel is assigned a specific wavelength within the optical spectrum. Multiple wavelengths can be used simultaneously on the same optical communication system such as the VLC system.
– Types of WDM:
  • Coarse WDM: in coarse WDM, the spacing between wavelengths is relatively large, typically in the range of 20 nm. This allows for a lower-cost implementation and is suitable for applications with lower data capacity requirements.
  • Dense WDM: dense WDM uses much narrower wavelength spacing, typically in the range of 0.8 nm to 0.4 nm. This enables a higher number of channels and greater data capacity, making dense WDM suitable for long-distance and high-capacity optical networks.
– Multiplexing and transmission: multiple signals, each operating at a different wavelength, are combined (multiplexed) into an optical path for transmission. This composite signal, containing multiple wavelengths, is transmitted over the spatial path simultaneously. At the receiving end, the signals are demultiplexed, and individual wavelengths are separated for further processing (see Fig. 6.2).

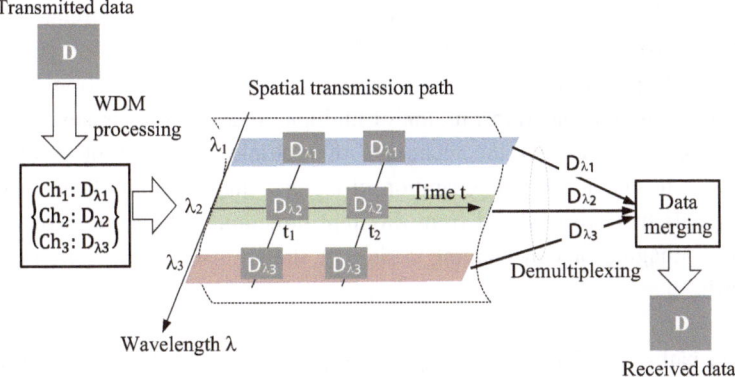

Fig. 6.2 Principle of WDM

Fig. 6.3 Example of duplexing link by using WDM

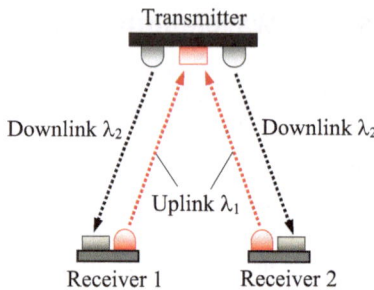

– Increased bandwidth: WDM significantly increases the bandwidth of VLC systems by allowing multiple signals to be transmitted in parallel. This is especially important for high-capacity applications, where the demand for data transmission capacity is substantial.

WDM has played a crucial role in meeting the ever-increasing demand for data transmission capacity in VLC systems. In addition, WDM can also be used for duplexing the uplinks and downlinks, i.e., all uplinks would employ wavelength λ_1, while all downlinks would use a second wavelength λ_2, as shown in Fig. 6.3. The main drawback of this approach is that it would not permit direct communication between the portables, unless each one was equipped with a second transmitter and/or receiver.

6.2   Space Division Multiplexing (SDM)

SDM is an important concept in modern optical communication systems, used to increase data capacity by exploiting the spatial dimension of the communication channel. Instead of dividing the channel based on the physical parameters such as time, frequency, or wavelength, SDM involves using different spatially separated communication paths or modes within the same optical medium space to support multiple communication channels. This allows for the simultaneous transmission of multiple, independent signals.

SDM involves the use of the technique of multiple-input-multiple-output (MIMO) configurations to enable spatial separation of signals, and an angle-diversity receiver to separate signals that are received from different directions. The angle-diversity receiver can be of an imaging or non-imaging design, as shown in Fig. 6.4. Here are key points about SDM:

– Multiple paths: in SDM, multiple physical paths or channels are established within the same communication medium. Each path is spatially separated from the others, often using different optical antennas, the angle-diversity receiver, or other spatial components.

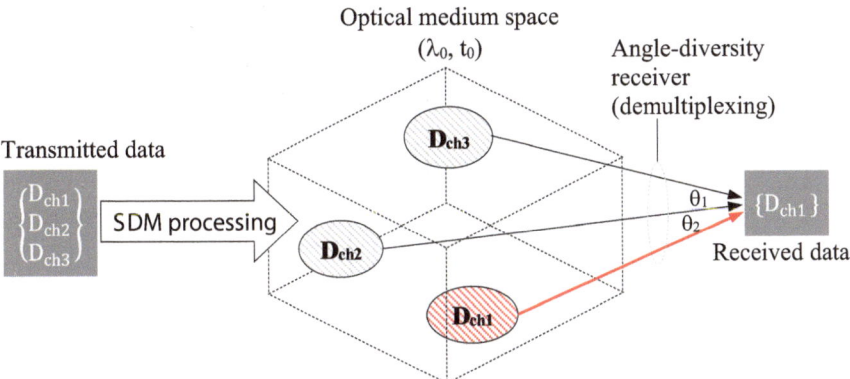

Fig. 6.4 Principle of SDM

– Parallel transmission: SDM enables parallel transmission of multiple signals, with each signal occupying a distinct spatial path. This is in contrast to WDM, TDM, and FDM, which divide the channel based on wavelength, time or frequency, respectively.
– Spatial separation methods:
  • Multiple optical antennas: in VLC systems, SDM can be implemented using multiple optical antennas, i.e., sending/receiving optical systems. Each antenna serves as an independent spatial path for signal transmission and reception.
  • . Multiple spatial channels: SDM can be achieved by using multiple channels, each serving as a separate spatial path for transmitting data.
  • . Optical MIMO: optical MIMO is a technology that utilizes multiple optical antennas at both the transmitter and receiver to create multiple spatial optical paths, improving data throughput and reliability in optical wireless communication.
– Advantages of SDM:
  • Increased capacity: SDM allows for a significant increase in the overall data capacity of a communication channel by exploiting spatial diversity.
  • Reduced interference: Spatially separated paths can help mitigate interference between signals, leading to improved signal quality and reliability.
  • Enhanced throughput: SDM is particularly useful in scenarios where traditional multiplexing techniques may be limited by bandwidth constraints.

As an example of SDM, consider a hub capable of establishing simultaneous, independent LOS links with several portable transceivers, as shown in Fig. 6.5. The hub can employ an angle-diversity receiver to minimize co-channel interference between different inbound receptions. in another potential application of SDM, one might construct a multiple-access LAN using quasi-diffuse transmitters and an imaging angle-diversity receiver.

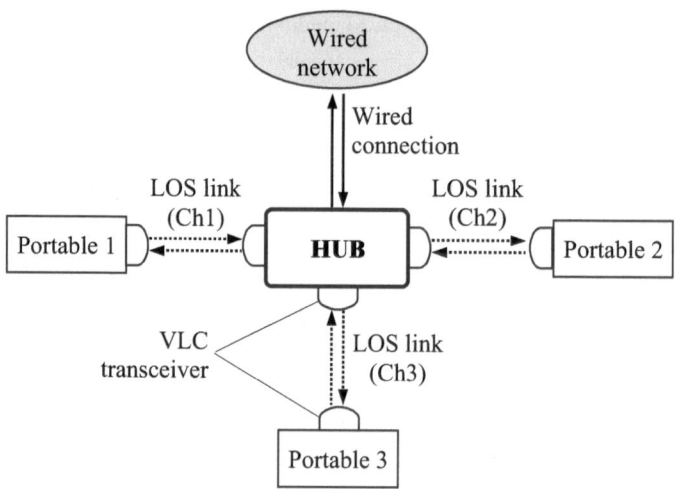

Fig. 6.5 Example of a hub using SDM technique to interconnect multiple portable devices

6.3   Polarization Division Multiplexing (PDM)

PDM technique can be also used for VLC systems. It allows two channels of information be transmitted on the same carrier frequency or wavelength by using optical waves of two orthogonal polarization states, as shown in Fig. 6.6.

In PDM, each signal is assigned a specific polarization state, and these signals can coexist on the same optical path without interference. Here are the key aspects of PDM:

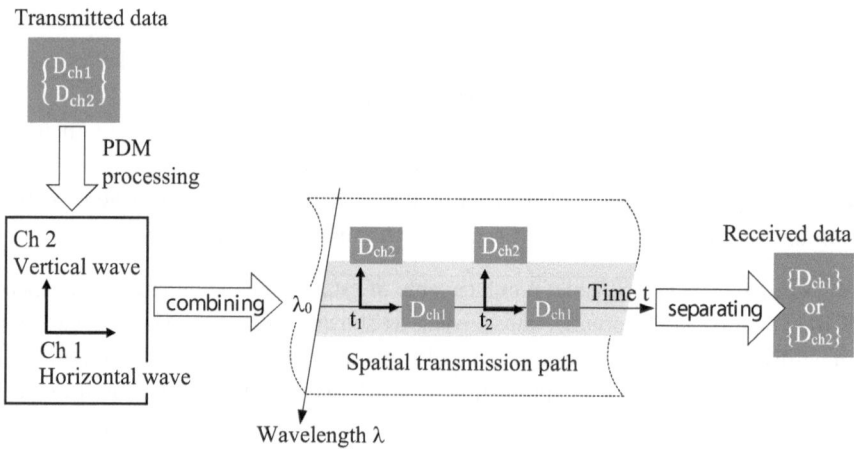

Fig. 6.6 Principle of PDM

- Polarization states: light is an electromagnetic wave that vibrates in different directions perpendicular to its propagation path. Polarization refers to the orientation of the electric field vector of the light wave. PDM utilizes different polarization states to distinguish between multiple signals.
- Orthogonal polarizations: in PDM, signals are transmitted using orthogonal polarization states. Orthogonal polarization states are mutually perpendicular, allowing them to be easily separated at the receiver.
- Transmission and receiving: the transmitter modulates each signal onto light waves with specific polarization states. These polarized signals are then combined and transmitted over the optical medium. At the receiving end, the signals are separated based on their polarization states using polarization-sensitive components.
- Components of PDM:
  - Polarization beam splitter: this optical component separates light into its orthogonal polarization components. It can be used to combine and separate polarized signals.
  - Polarization maintaining device: this device is designed to maintain the polarization state of light, ensuring that the different polarized signals remain distinguishable during transmission.
  - Polarization controller: this device can adjust the polarization state of light, ensuring that transmitted signals are properly aligned for reception.
- Advantages of PDM:
  - Increased data capacity: PDM allows for an increase in the data-carrying capacity of optical communication systems by transmitting multiple signals in orthogonal polarization states.
  - Improved spectral efficiency: PDM improves spectral efficiency by allowing multiple signals to share the same frequency band without interfering with each other.
  - Enhanced performance: PDM is less susceptible to some types of impairments, such as polarization mode dispersion, which can affect traditional communication systems.
- Applications of PDM:
  - PDM is typically used together with phase modulation or optical quadrature amplitude modulation (QAM), such as polarization-multiplexed quadrature amplitude modulation (PM-QAM) [2], allowing transmission speeds of 100 Gbit/s or more over a single wavelength. Sets of PDM wavelength signals can then be carried over WDM infrastructure, potentially substantially expanding its capacity. Multiple polarization signals can be combined to form new states of polarization, which is known as parallel polarization state generation.
  - Differential cross-polarized wireless communications is a novel method for polarized antenna transmission utilizing a differential technique [3]. Figure 6.7 shows a schematic of cross-polarized system. The method uses a linear cross-polarized

Fig. 6.7 Schematic of the cross-polarized system

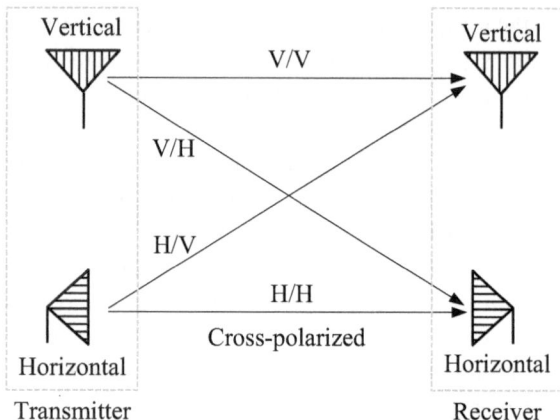

antenna, for example, a vertical and horizontal polarized antenna, improving performance and power efficiency of wireless communications. This solution is simple, compact and does not require any bandwidth expansion.

6.4   Optical Time Division Multiplexing (OTDM)

Nature of time division multiplexing (TDM) is a technique used in telecommunications to transmit multiple signals over a single communication channel by dividing the channel into discrete time slots. Each time slot is then assigned to a specific signal or user, allowing multiple signals to share the same transmission medium without interfering with each other.

TDM in optical communication systems is similar to traditional TDM but is applied in the optical domain [4, 4]. In optical TDM (OTDM), optical signal of the same wavelength is divided and become multiple signals by time and allocates multiple channels, each channel is assigned specific time slots during which they can transmit their optical signals, as shown in Fig. 6.8.

Here is a basic overview of how TDM works:

– Time Slots: the communication path is divided into fixed time intervals, known as time slots. Each time slot corresponds to a brief, fixed period during which a particular channel can transmit data.
– Assignment of time slots: different communication paths are assigned specific time slots in a cyclic or sequential manner. For example, if there are three signals, the first signal may use the first time slot, i.e., channel 1, the second signal the second time slot, i.e., channel 2, and so on. After the last signal has transmitted, the cycle repeats.

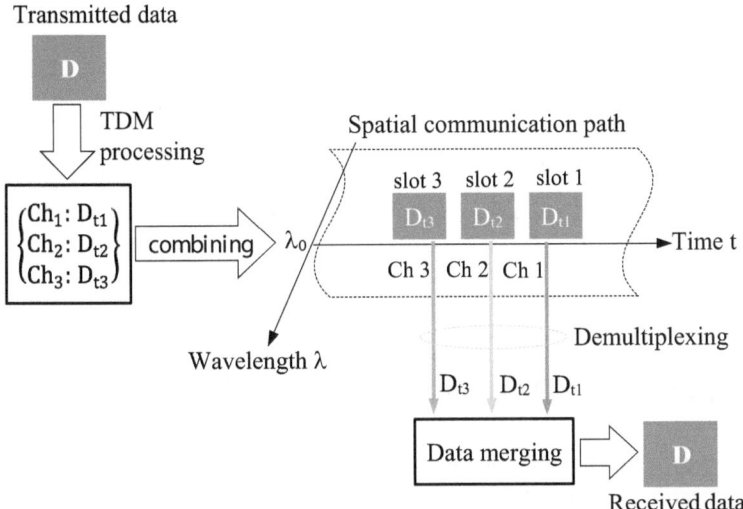

Fig. 6.8 Principle of OTDM

- Synchronization: data streams sharing the communication path must be synchronized to the time slots. This ensures that each data stream knows when it is their turn to transmit and when to expect incoming data.
- Multiplexing and transmission: during their assigned time slots, each signal transmits their data. The signals are combined (multiplexed) into a composite signal for transmission over the communication path. At the receiving end, the composite signal is demultiplexed, and the individual signals are extracted based on the assigned time slots.

One of the advantages of TDM is its simplicity and efficiency in utilizing available bandwidth. However, it may face challenges in situations where the data rates of different signals vary significantly or when the number of data streams is large, as it requires precise synchronization among all data streams. OTDM is often used in conjunction with WDM to achieve both time and wavelength multiplexing (Fig. 6.9), thereby to achieve higher speed and large-capacity optical digital transmissions.

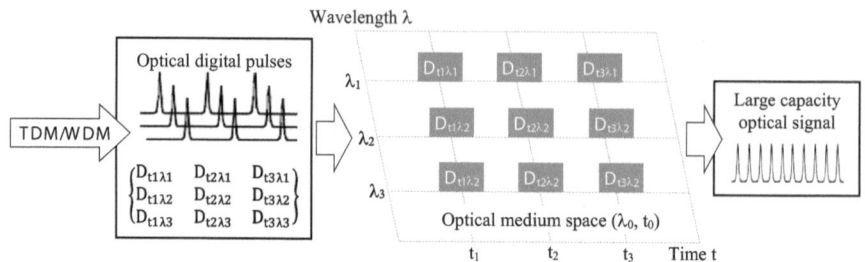

Fig. 6.9 Conjunction of OTDM and WDM

6.5   Optical Code Division Multiplexing (OCDM)

Nature of code division multiplexing (CDM) is a digital cellular technology used in telecommunications. Unlike TDM and FDMA, which divide the channel into time slots or frequency bands, CDM uses a unique code to distinguish each channel's signal. Different codes employ different orthogonal or quasi orthogonal sequences, permitting them to transmit simultaneously. The transmitted signals are spread over a wide bandwidth, and the receiver uses the corresponding code to demodulate the specific signal of interest, as shown in Fig. 6.10.

The power efficiency achieved by CDM varies, depending upon the duty cycle of the transmitted waveforms. In telecommunications, CDM is widely used in modern digital cellular networks, such as WCDMA (Wideband Code Division Multiple Access) [5].

CDM techniques can also be extended to optical communication systems [6, 7]. In Optical CDM (OCDM), optical signals also are assigned unique codes, and they are spread over the optical spectrum to process and send using these codes. On the receiving side,

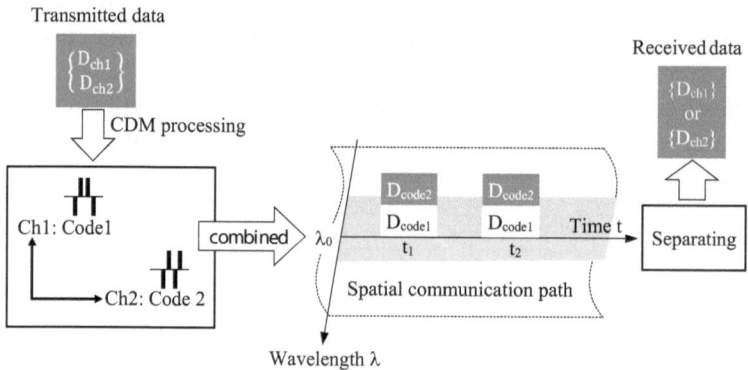

Fig. 6.10 Principle of CDM

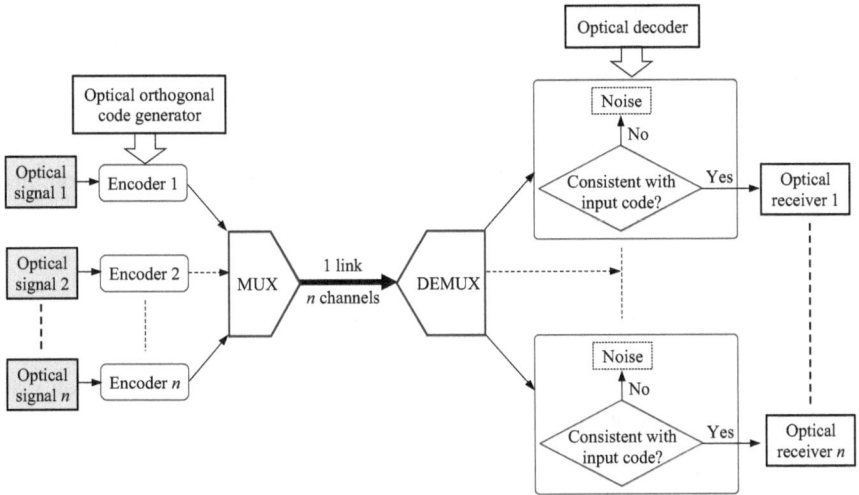

Fig. 6.11 Schematic of OCDM method

the receiver, equipped with the corresponding decoding mechanism, can isolate or recover the signals of individual channels, i.e., the original information signal can be obtained by processing the transmitted signal with the same code, but signals processed with other codes become noise and the original information signal cannot be restored. The schematic of OCDM method as shown in Fig. 6.11.

Key features of CDM include:

- Unique codes: in CDM, each channel is assigned a unique code. This code is used to modulate the signal in the channel before transmission. The unique codes enable multiple channels to share the same frequency or wavelength band without interference.
- Spread spectrum technology: CDM employs spread spectrum technology, which spreads the signal over a wide frequency band. This spreading helps in distinguishing between different channels and provides a level of security against interference.
- Simultaneous transmission: multiple channels can transmit and receive data simultaneously on the same frequency, as long as their unique codes are different. This allows for efficient use of available bandwidth and increased capacity in the communication system.
- Soft handoff: CDM supports *soft handoff*, which allows a mobile device to be in communication with multiple base stations simultaneously.
- Interference tolerance: CDM is known for its tolerance to interference and multipath fading, making it suitable for wireless communication environments.

In the past, CDM technology has been widely used in 2G and 3G mobile communication systems. Also, hybrid OCDM and WDM have flourished as successful schemes for expanding the transmission capacity as well as enhancing the security for OCDMA [8]. However, with the evolution of wireless technologies, such as 4G LTE (Long Term Evolution) and 5G, other access technologies like OFDMA have become more prevalent. Nevertheless, CDM remains a fundamental concept in wireless communication, and its principles have influenced the development of subsequent wireless standards.

6.6   Optical Multilevel Modulation (OMLM)

Multilevel Modulation (MLM) is a modulation technique used in communication systems to transmit multiple bits of information per symbol by employing more than two levels or states. The fundamental idea behind multilevel modulation is to increase the data rate by transmitting multiple bits in each symbol period. Figure 6.12 shows an example for comparison a binary and a three-level modulation. In common binary modulation, each symbol only represents a binary bit (0 or 1), in contrast, in MLM, each symbol can represent more than two levels, allowing the transmission of multiple bits in each symbol.

Optical MLM (OMLM) is the use of the MLM technique for modulating optical signals and has been widely used in optical communication [10, 11] and VLC [12, 13] systems for different purposes. In Ref.13, authors proposed an OMLM scheme for VLC system with mobile phone camera using the overlapping of two LED sources, and the two light sources are modulated by an OOK and a Manchester signal respectively, as shown in Fig. 6.13a. The experimental results demonstrate that the proposed OMLM scheme can achieve a net data rate of 4.32 Kbit/s. Authors of Ref.14 proposed a high-accuracy VLC-based indoor positioning system using OMLM. In their system, Each LED source transmits different IDs (00, 11, 10, 01) related to its own physical location by the MLM (5-level). At the transmitter, binary data are mapped from the serial data frame to parallel data with the five amplitude levels. The first four levels (level 0-level 3) are used for pulse-amplitude

Fig. 6.12 Binary modulation versus MLM

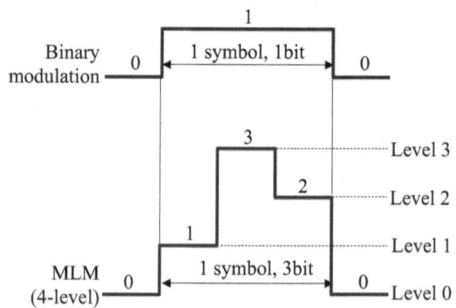

modulation (PAM) data transmission and the last level (level-4) for data detection, as shown in Fig. 6.13b.

MLM technique plays a crucial role in modern communication systems. However, it is worth noting that while MLM increases the data rate by transmitting more bits per symbol, it may also make the system more susceptible to noise and interference. The

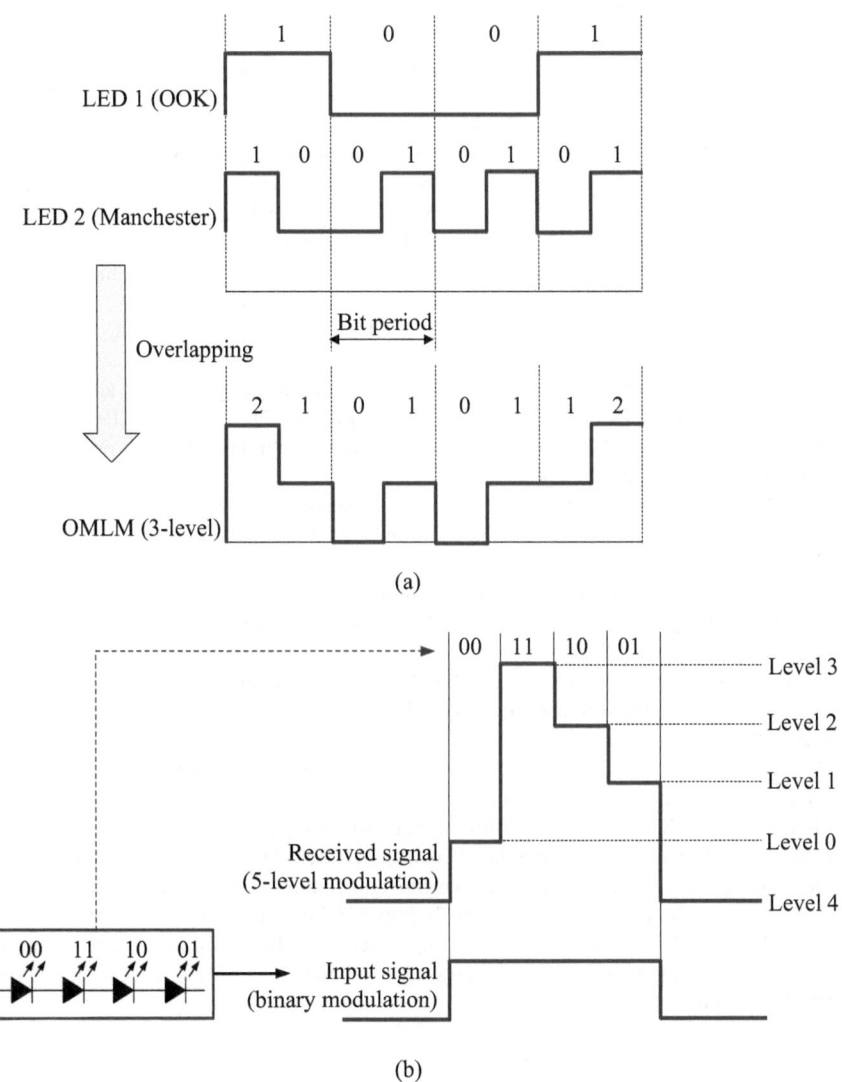

Fig. 6.13 Examples of OMLM's applications: **a** OMLM using the overlapping of two light sources and **b** OMLM for positioning

trade-off between data rate and system robustness needs to be carefully considered in the design of communication systems.

References

1. J.M. Kahn, J.R. Barry: Wireless infrared communications, Proc. of the IEEE **85**(2), 256–298 (1997)
2. Y. Wang, K. Kasai, M. Nakazawa: Polarization-multiplexed, 10G symbol/s, 64QAM coherent transmission over 150 km with OPLL-based homodyne detection employing narrow linewidth LDs, IEICE Electro. Express, **8**(17), 1444–1449 (2011)
3. S. Ghadimi: Differential cross-polarized wireless communications, Wireless Eng. and Tech., **10**(2), 34–40 (2019) https://doi.org/10.4236/wet.2019.102003
4. Y. Yamabayashi, H. Toba, M. Nakazawa: State-of-the-art and future perspectives of time division multiplexing (TDM) high bit rate optical transmission, Laser Rev., **27**(4), 231–239 (1999)
5. M. Nakazawa: Advanced photonic technology for the research of optical time-division multiplexing and its future prospects, J. IEICE, **102**(8), 809–814 (2019)
6. M. Sawahashi, T. Nakamura, K. Higuchi: W-CDMA technique, NTT DoCoMo Tech. Jour., **18**(3), 56–69 (2010)
7. K. Kitayama, H. Sotobayashi: Optical code division multiplexing (OCDM) and its application to photonic network, IEICE Trans. Fundamentals, **E82-A**(12), 2616–2626 (1999)
8. H. Sotobayashi: Optical code division multiplexing and ultrafast photonic processing, Japan. Jour. of Opt., **32**(1), 26–31 (2003)
9. Isaac A. M. Ashour, S. Shaari, P. S. Menon, Hesham A. Bakarman: Optical code-division multiple-access and wavelength division multiplexing: hybrid scheme review, Jour. of Computer Science, **8**(10), 1718–1729 (2012)
10. X. Zhou, J. Yu: Multi-level, multi-dimensional coding for high-speed and high-spectral-efficiency optical transmission, Jour. of Lightwave Tech., **27**(16), 3641–3653 (2009), https://doi.org/10.1109/JLT.2009.2022765
11. L. Combi, A. Gatto, P. Boffi, U. Spagnolini, P. Parolari: Optical multilevel pulse width modulation for analog mobile fronthaul, Photonics **2018**, 5, 49 (2018), https://doi.org/10.3390/photonics5040049
12. H. S. Jeong, J. H. Cho, H. K. Sung: Evaluation of performance enhancement of optical multi-level modulation based on direct modulation of optically injection-locked semiconductor lasers, Photonics 2021, **8**(4) 130 (2021), https://doi.org/10.3390/photonics8040130
13. J. Shi, J. He, R. Deng, Y. Wei, F. Long, Y. Cheng, L. Chen: Multilevel modulation scheme using the overlapping of two light sources for visible light communication with mobile phone camera, Opt. Express, **25**(14), 15905–15912 (2017), https://doi.org/10.1364/OE.25.015905
14. N. Q. Pham, V. P. Rachim, W. Y. Chung: High-accuracy VLC-based indoor positioning system using multi-level modulation, Opt. Express, **27**(5), 7568–7584 (2019), https://doi.org/10.1364/OE.27.007568

VLC Standards

7

Standardization can both avoid mutual interference between different VLC products and hold good compatibility between different VLC applications, which is an effective way to ensure the healthy development of VLC technology. This chapter introduces three representative current VLC standards respectively: JEITA CP-122X (X = 1, 2, 3) for low-speed unidirectional ID communications, IEEE 802.15.7 for medium to high-speed short-range communications, and ITU-T G.9991 (a.k.a. G.VLC) for VLC transceiver of indoor high-speed optical networks, such as a Li-Fi system.

Standardization is an important and effective way to ensure the development of VLC technology and markets. This is because with the development of VLC technology, it is expected that a variety of VLC-related applications and products will emerge. Standardization can both avoid mutual interference between different VLC products and prevent the adverse effects of VLC equipment on existing infrared equipment, as well as ensure interoperability in the use of different communication methods and compatibility between different VLC applications within a designated VLC service area. VLC standards can also offer valuable support to VLC-based IoT, a landscape composed of a very diverse set of solution providers, without standards it has been impossible to create a shared ecosystem.

TC34 (IEC Technical Committee: Lamps and related equipment) has developed a standard for illumination that covers the connection between lamps and power supplies. VLC standards are various related protocols between senders and receivers. JEITA CP-1221 for VLC system (March 2007) [1], JEITA CP-1222 for VL-ID system (June 2007) [2], and JEITA CP-1223 for VL beacon system (May 2013) [3] are early VLC standards formulated by the VLCC (VLC Consortium) of Japan for low-speed data transmission with a data rate of only 4.8 Kbps. In September 2011, IEEE 802.15.7 established a standard for

© The Author(s), under exclusive license to Springer Nature Switzerland AG 2025   93
X. Lin, *Visible Light Communications*, Synthesis Lectures on Communications,
https://doi.org/10.1007/978-3-031-64475-7_7

short-range VLC [4] with a data rate of up to 96 Mbps. ITU-T (International Telecommunication Union Telecommunication Standardization Sector) G.9991 is a high-speed VLC standard developed by ITU-T for indoor LOS optical networks, which was approved in March 2019 [5] and updated to the latest version in April 2021 [6].

7.1 JEITA for VLC

In order to speed up the research, development, and commercialisation of VLC techniques, the VLCC, which is a predecessor of VLCA (VLC Association) was formed in Japan 2003. In 2006, members of VLCC proposed two standards that as JEITA CP-1221 and JEITA CP-1222 for VLC systems and VL ID systems, and these two standards were accepted by JEITA in March 2007 and June 2007, respectively. In 2013, members of VLCC worked together again to introduce a VL beacon system that as JEITA CP-1223, it is a simplified and improved version of CP-1222, and it was accepted by JEITA in May 2013.

7.1.1 JEITA CP-1221: VLC System

JEITA CP-1221: VLC System (hereinafter referred to as CP-1221) is the most basic standard for VLC systems. This standard has two purposes, one is that to present an indicator minimum in order to prevent the interference between different optical communication equipment, and other is that to define a minimum necessary requirement in various VLC applications. Here are the key aspects of CP-1221:

- Scope of application: CP-1221 is applicable to optical wireless communication (OWC) systems using VL wave as carrier, and it divides the system into two main parts: the physical layer and the application and upper layers, as shown in Fig. 7.1. CP-1221 makes relevant provisions for a part of the physical layer, which includes the transmitting (Tx) and receiving (Rx) devices as well as their interfaces with spatial channels.
- Wavelength range and its utilization: in CP-1221, the wavelength range for VLC is defined to be 380–780 nm. CP-1221 allows each VLC application can use arbitrary wavelength that minimum wavelength unit of 1 nm within the above range.
- Communication schemes: CP-1221 allows the modulation of digital or analog signals by using both BIM (Sect. 5.1) and SCM (Sect. 5.3) schemes. When SCM scheme is used, the frequency allocation of subcarrier is shown in Fig. 7.2. In Fig. 7.2, the frequency range 1 from 15 to 40 kHz is used for VL-tag systems, and the range 3 that is greater than 1 MHz is used for the application that needs higher speed communication, such as VL-multimedia systems. In the range 2 from 40 kHz to 1 MHz, because the

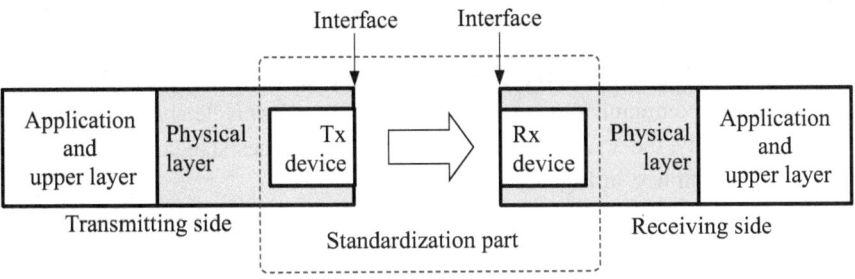

Fig. 7.1 Scope of application

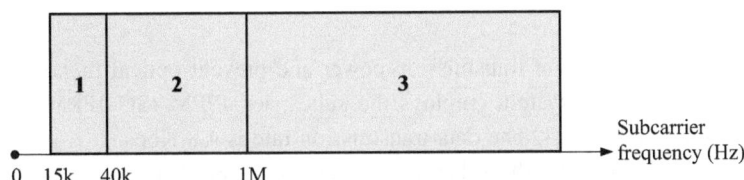

Fig. 7.2 Frequency allocation of subcarrier

noise from fluorescent lamps is quite large (Sect. 4.3.1), it is not appropriate to use this range for VLCs, especially indoor VLC systems. Specific application undecided.

– Components: in CP-1221, the main components are the light-emitting and the light-receiving element, both they can have a single or an array type. CP-1221 also assumes that the light-emitting element is a device with a diffuse spatial radiation distribution like to that of illumination light.

– Transmission distance and communication range: the necessary communication quality must be ensured in both transmission distances and communication ranges.

– Safety: all VLC systems should comply with the following 3 standards for human safety, i.e., 1. IEC 60,825–1 for *Safety of laser products-Part 1: Equipment classification and requirements*, 2. IEC 62,471 for *Photobiological safety of lamps and lamp systems*, and 3. JIS (Japanese Industrial Standards) C 7550 (old: JIS TS C 0038) for *Photobiological safety assessment service.*

– Compatibility and prevention of interference: to ensure compatibility between VLC devices, standardized measurement and evaluation methods should be applied to each device. Additionally, in order to avoid jamming with existing infrared devices and interference with background light noise such as sunlight and fluorescent lamps, CP-1221 recommends that VLC devices should be equipped with appropriate optical or electrical filters according to the use environment, and comply with EMC (Electromagnetic compatibility) standards.

7.1.2   JEITA CP-1222: Visible Light ID System

JEITA CP-1222: VL ID System (hereinafter referred to as CP-1222) prescribes a unidirectional, one-to-one communication system using VL of peak wavelength of 380–780 nm as the medium, namely, the VL ID system. It specifically defines standards for the interface portion that is illustrated in Fig. 7.3.

In CP-1222, the interface protocol consists of a physical layer (L1), a frame layer (L2), an ID/data layer (L3) and an application layer (L4). CP-1222 aims to establish a unified standard for the common lower communication layers (L1, L2, L3) that are applicable to various uses of the VL ID system, promoting their shared utilization. It does not address the higher communication layer (L4) specific to individual applications.

Physical Layer

In order to maintain constant transmission power and prevent optical flicker during data transmission, the VL ID system employs the subcarrier 4PPM (SC-4PPM) modulation scheme as shown in Fig. 7.4. The data transmission rate is 4.8 Kbps.

The CP-1222 defines that the subcarrier frequency is 28.8 kHz, the transmission rate of the 4PPM signal is 4.8kbps, and the subcarrier wave and 4PPM signal are assumed to be phase-synchronized. In 4PPM modulation scheme, one symbol time (D) is divided equally into four slots (C), and transmitting 2 bits of data by each symbol (Fig. 5.2). So, in Fig. 7.4, the symbol time $D = 2 \times (1/4800) = 0.146$ ms, and the slot time $C = D/4 = 0.104$ ms.

Also, if each value a, b, and c (but $a < c$) of the waveform in Fig. 7.4 is defined, then the signal amplitude $= c-a$, modulation depth $= (c-a)/c$. By adjusting the values of a, b and c, the system can be optimized for various applications. For example, if prioritizing communication efficiency over brightness, setting $b = a = 0$ and turning the light source completely off during pulseless intervals, as shown in Fig. 7.5a. In this case, the light source is switched on or off with 100% modulation depth. On the other hand, for applications such as lighting fixtures where brightness is essential, setting $b = c$, as shown in Fig. 7.5b, increases the illumination time ratio and enhances luminous efficiency.

Fig. 7.3 VL ID system

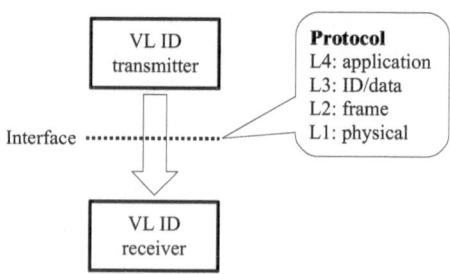

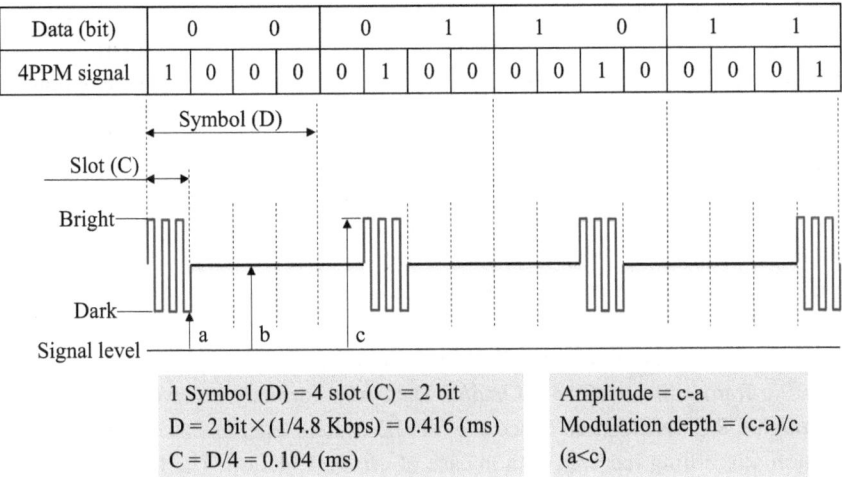

Fig. 7.4 SC-4PPM signal waveform for VL ID system

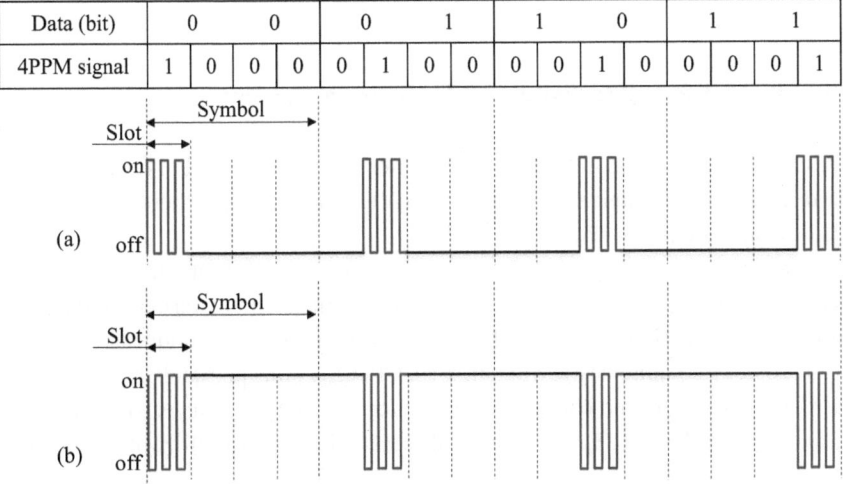

Fig. 7.5 SC-4PPM signal waveform with light source on/off: **a** b = a = 0 and **b** b = c

Frame Layer

The frame layer in CP-1222 has two structures, Type A and Type B. The respective frame structures for each type are shown in Table 7.1. Frames of each type consist of common elements: Start of Frame (SOF), Payload, and End of Frame (EOF). The Start of Frame includes a Preamble (PRE) of 6 bit and Frame Type (F-Type) of 8 bit, and in the case of

Table 7.1 The frame structure for VL ID system

SOF (start of frame)		Payload			EOF (end of frame)
		Type A	Type B		
PRE (preamble)	F-TYPE (frame type)	Data	ID	Data	CRC-16
6 bit	8 bit	512 bit	128 bit	384 bit	16 bit

Type B, the Payload field of 512 bit is divided into an ID section of 128 bit and a Data section of 384 bit.

The End of Frame is a CRC-16 (CRC: cyclic redundancy checks) of 16-bit to determine whether the frame information, ID and/or Data was correctly received on the receiving side through CRC verification. Since the VL ID system operates as unidirectional communication, discarding received data in case of errors detection. The CRC field stores the result of the CRC calculation. The CRC-16 generation polynomial is given by $X^{16} + X^{15} + X^2 + 1$, and in the binary case, $X = 2$; the CRC operation range contains F-TYPE and Payload. The idle time between frames, the transmission order of Type A and Type B and their mixing ratio for are freely configurable.

7.1.3   JEITA CP-1223: Visible Light Beacon System

JEITA CP-1223: VL Beacon System (hereinafter referred to as CP-1223) prescribes the single directional communication system with VL as the medium (hereinafter referred to as VL beacon system), as shown in Fig. 7.6.

VL beacon system is a system for providing various applications such as identification of matters, providing positional information, and establishment of various guiding systems by radiation transmission of simple information or ID information unique to the VL source from VL sources ubiquitously surrounding us. CP-1223 aims at establishment of a unified standard concerning lower communication layer, i.e., physical layer (L1), frame layer (L2), and ID/data layer (L3) common to these applications to utilize it commonly, and

Fig. 7.6 VL beacon system

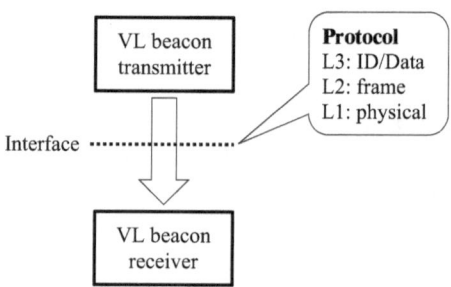

do not deal with upper communication layer which depends upon individual applications. So, CP-1223 especially prescribes the standard concerning interface part which consists of L1, L2, and L3 in the VL beacon system of Fig. 7.6.

Physical Layer
The VL beacon transmitter can transmit ID. The ID code system used is selectable, and various services, for example, equipment identification, position information and data transmission function can be provided. The receiver can obtain its positional information by ID resolution. Detection of peripheral services based on the present position, transmission position report function at the time of emergency report and so forth can be realized in buildings or underground malls where use of GPS (global positioning system) is difficult. Further, the function can be used as the means for a robot to detect its position.

Same as the CP-1222, carrier frequency of this system shall be VL of peak wavelength of 380–780 nm. VL is intensity modulated by I-4PPM (Inverted-4PPM) signals, of which information data rate is 4.8 Kbps, as shown in Fig. 7.7. The signal amplitude $= b-a$ and the modulation depth $= (b-a)/b$.

Also same as the CP-1222, in Fig. 7.7, the symbol time $D = 2 \times (1/4800) = 0.146$ ms, and the slot time $C = D/4 = 0.104$ ms.

Frame Layer
Frame structure is shown in Table 7.2. Same as the CP-1222, the frame in the CP-1223 consists of start part (SOF), payload, and end part (EOF). Further, SOF is divided into preamble (PRE) and Frame-type (F-Type), and information part, i.e., payload consists of ID and/or DATA part. The EOF is CRC-16.

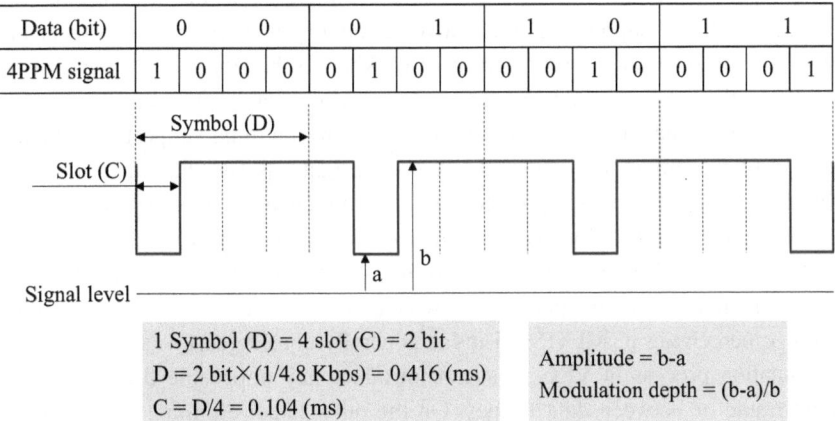

Fig. 7.7 I-4PPM signal waveform for VL beacon system

Table 7.2 The frame structure for VL beacon system

SOF (start of frame)		Payload	EOF (end of frame)
PRE (preamble)	F-TYPE (frame-type)	ID/data	CRC-16
6 bit	8 bit	128 bit	16 bit

The PRE of 6 bit is set as the frame start position. In the F-type, the codes indicating kinds of payload are represented by 8 bit in order to enable use of IDs and codes of different plural system and transmission of data. The Payload consists of 128 bit ID and/ or data. CRC field length shall be 16 bit. Reception side judges whether frame data were correctly received. Since VL beacon system is single directional communication incapable of re-transmission request, received data are discarded if error was detected. Calculation results of CRC are stored in CRC field. Calculation ranges of CRC are F-type and payload, and generation polynomial is given by $X^{16} + X^{15} + X^2 + 1$, where $X = 2$ in the binary case.

7.2   IEEE Standard for VLC

IEEE 802.15.7: Short-Range Wireless Optical Communication Using Visible Light (here-inafter referred to as IEEE 802.15.7) defines a PHY (physical) and MAC (medium access control) layer for short-range optical wireless communications using VL spectrum from 380 to 780 nm. IEEE 802.15.7 is capable of delivering data rate up to 96 Mbps suffi-cient to support audio and video multimedia services, and also considers mobility of the visible link, compatibility with VL infrastructures, impairments due to noise and interfer-ence. IEEE 802.15.7 supports multiple diverse topologies, such as peer-to-peer and star topologies, with data rates ranging from 11.67 kbps to 96 Mbps for indoor and outdoor applications. IEEE 802.15.7 adheres to applicable eye safety regulations.

The two main challenges for LED-based VLC are dimming support and flicker miti-gation [7]. Flicker refers to the fluctuation of the brightness of light. Any potential flicker resulting from modulating the light sources for communication must be mitigated because flicker can cause harmful physiological changes in humans. To avoid flicker, the changes in brightness must fall within the maximum flickering time period (MFTP). The MFTP is defined as the maximum time period over which the light intensity can change without the human eye perceiving it, MFTP < 5 ms is generally considered eye safe [8]. Therefore, the modulation process in VLC must not introduce any noticeable flicker either during the data frame or between data frames. On the other hand, dimming support is another important consideration for LED-based VLC for power savings and energy efficiency for light sources. It is desirable to maintain communication while a user arbitrarily dims the light source.

Table 7.3 Operating modes of three PHY types

	PHY I	PHY II	PHY III
Data rate	11.67−266.6 kbps	1.25−96 Mbps	12−96 Mbps
Light source type	Single	Single	Multiple
Modulation	OOK, VPPM	OOK, VPPM	CSK
FEC	RS, CC	RS	RS
RLL code	Manchester (OOK) 4B6B (VPPM)	8B10B (OOK) 4B6B (VPPM)	None
OCR	$\leq$400 kHz	$\leq$120 MHz	$\leq$24 MHz

7.2.1 Physical Types for VLC

IEEE 802.15.7 offers three physical (PHY) types for VLC. As shown in Table 7.3, PHY I operates from 11.67 to 266.6 kbps, PHY II operates from 1.25 to 96 Mbps and PHY III operates between 12 and 96 Mbps. PHY I and PHY II are defined for a single light source, and support OOK and variable PPM (VPPM) schemes. PHY III uses multiple optical sources with different wavelengths (colors) and uses CSK modulation scheme. The different modulation schemes allow trade-offs between data rates and different dimming ranges.

Each PHY mode contains mechanisms for modulating the light source and channel coding for forward error correction (FEC). IEEE 802.15.7 supports various FEC schemes such as Reed-Solomon (RS) and convolutional codes (CC). There is also a run length limited (RLL) line coding in PHY I and II to avoid long runs of "1" and "0" that could potentially cause flicker. Various RLL line codes such as Manchester and 8B10B for OOK, and 4B6B for VPPM are defined in IEEE 802.15.7. Each PHY modulation mode has an associated optical clock rate (OCR). OCR for PHY I is chosen to be $\leq$400 kHz for slow-speed LEDs such as traffic lights; for PHY II, OCR is chosen to be $\leq$120 MHz to accommodate fast LEDs used in mobile and portable devices for communication; and for PHY III, OCR is chosen to be $\leq$24 MHz, which is the maximum clock rate supported by current-infrastructure white LEDs.

7.2.2 Dimming Support Using Various Modulation Schemes

Dimming Using OOK

OOK modulation is the simplest modulation scheme for VLC (Sect. 5.1.1), where the LEDs are turned on or off dependent on the data bits being 1 or 0. While the modulation is logically OOK, OOK *off* does not necessarily mean the light is completely turned off;

rather, the intensity of the light may simply be reduced as long as one can distinguish clearly between the *on* and *off* levels and achieve dimming.

In IEEE 802.15.7, OOK dimming can be achieved by two methods: one is redefining the *on* or *off* levels of the OOK symbol to have a lower intensity, and other is the levels can remain unchanged, the average duty cycle of the waveform can be changed by the insertion of *compensation* time into the modulation waveform. The compensation time is realized by fully turning on or off the light source for the required duration to provide dimming.

The frame structure for OOK dimming is shown in Fig. 7.8, it consists of a preamble for synchronization, a PHY header that provides details on the frame such as the frame length, modulation, and coding, and the data payload frame. When compensation time is added, it is possible for the receiver to lose synchronization for long compensation times. Hence, in this dimming frame structure, the data frame is broken into subframes, and each subframe can be preceded by a resynchronization (resync) field using a 1010... maximum transition sequence pattern that aids in readjusting the data clock after the compensation time. In Fig. 7.8, short sync fields, i.e., 1010 patterns are used to resync the receiver before the data subframes.

Dimming Using VPPM

The PPM as one of important modulation scheme for VLC has been described in Sect. 5.1.3. The duty cycle of each symbol in PPM is a certain amount of time. Compared to the PPM, VPPM (VPPM) encode bits by changes the duty cycle of each symbol. In VPPM, the variable term represents the change in the duty cycle response to the requested dimming level.

Figure 7.9 is basic concept of VPPM. VPPM is similar to 2-PPM when the duty cycle is 50% visibility as shown in Fig. 7.9a, and the duty cycle is adapted using PWM (Sect. 5.1.4) for other visibility levels, as shown in Fig. 7.9b, the pulse width ratio (b/a) of PPM can be adjusted to produce the required duty cycle for supporting dimming.

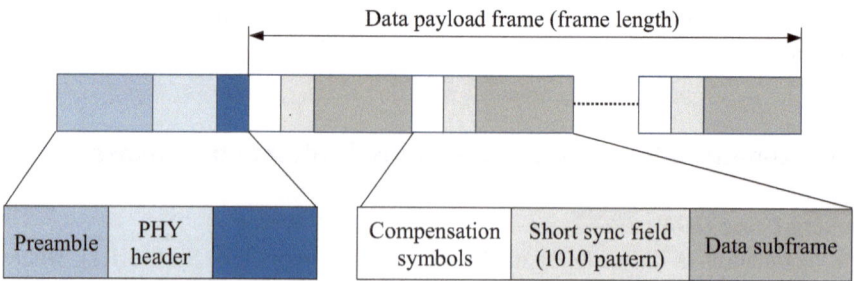

Fig. 7.8 The frame structure for OOK dimming

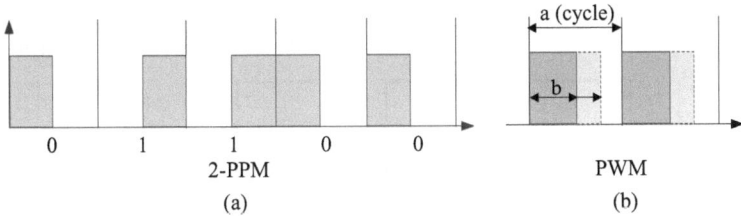

Fig. 7.9 Basic concept of VPPM: **a** 2-PPM, and **b** PWM

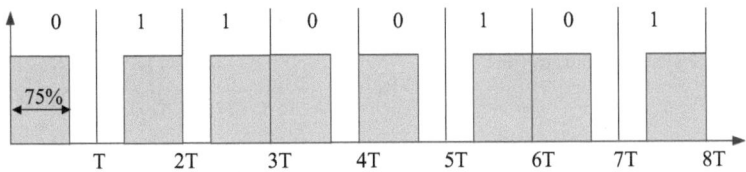

Fig. 7.10 Waveform of VPPM scheme with 75% pulse width (duty cycle)

Figure 7.10 shows an example waveform of how VPPM can attain a 75% dimming duty cycle requirement, where both logic 0 and logic 1 have a 75% pulse width.

Dimming Using CSK

The use of multicolor LEDs (Fig. 2.5 in Sect. 2.1.4) forms the principle behind CSK (Sect. 5.4) modulation. CSK modulation is similar to frequency shift keying in that the bit patterns are encoded to color (wavelength) combinations. For example, for 4-CSK (two bits per symbol) the light source is wavelength keyed such that one of four possible wavelengths (colors) is transmitted per bit pair combination. In order to define various colors for communication, the IEEE 802.15.7 standard breaks the spectrum into 7 color bands (Fig. 5.13 in Sect. 5.4) in order to provide support for multiple LED color choices for communication.

Figure 7.11 shows the CSK system configuration for PHY III with three color (bands i, j, and k) light sources. After scrambling and channel coding, the logical data values of 0 and 1 are transformed into xy values, according to a mapping rule on the xy color coordinates by the color coding block. The scrambler is necessary to create pseudo-random data and prevent data-pattern-dependent color shifts. These xy values are transformed into intensity P_i, P_j, and P_k. The relation between the coordinates and the intensity is given by

$$x_p = P_i x_i + P_j x_j + P_k x_k, \tag{7.1}$$

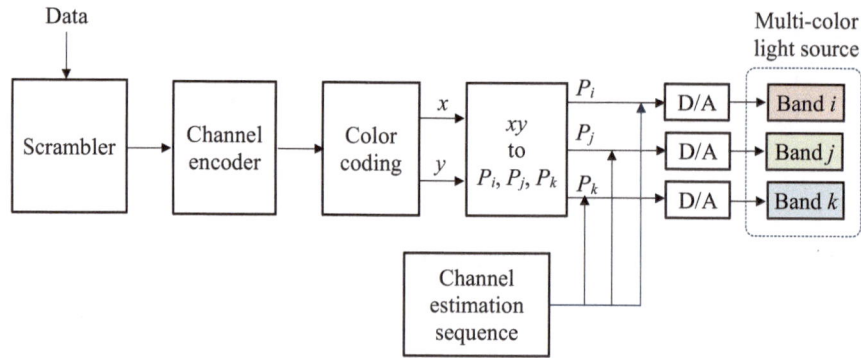

Fig. 7.11 CSK system diagram for PHY III

$$y_p = P_i y_i + P_j y_j + P_k y_k, \tag{7.2}$$

and

$$P_i + P_j + P_k = 1. \tag{7.3}$$

The total power of all CSK light sources is constant, although each light source may have a different instantaneous output power. CSK dimming ensures that the average optical power from the light sources is kept constant and maintains the requisite intensity of the center color of the color constellation. Thus, there is no flicker issue associated with CSK due to amplitude variations. CSK dimming employs amplitude dimming and controls the brightness by changing the current driving the light source. Also, CSK supports amplitude changes with D/A (digital-to-analog) converters, thus allowing higher order modulation support to provide higher data rates at a lower optical clock frequency. On the receiver side, xy values are calculated from the received light powers of 3 colors, and xy values are decoded into the received data.

7.2.3 Methods of Flicker Mitigation

The flicker in VLC is classified into two categories according to its generation mechanism: intra-frame flicker and inter-frame flicker. Intraframe flicker is defined as the perceivable brightness fluctuation within a frame. Inter-frame flicker is defined as the perceivable brightness fluctuation between adjacent frame transmissions [7]. A summary of the different mitigation techniques for inter-frame and intra-frame flicker is provided in Table 7.4.

Table 7.4 Flicker mitigation methods

	Intra-frame flicker mitigation (data transmission)	Inter-frame flicker mitigation (receive or idle)
OOK	– Dimmed OOK mode – Using RLL code	Using idle patterns
VPPM	– No inherent intra-frame flicker – Using RLL code	
CSK	– Constant average power across multiple light sources – Scrambler – High OCR with MHz	

Intra-Frame Flicker Mitigation

Intra-frame flicker mitigation refers to mitigation flicker within a *data transmission* frame. As shown in Table 7.4, intra-frame flicker in OOK is avoided by using the dimmed OOK mode and RLL coding; VPPM uses RLL code and does not cause any inherent inter-frame flicker; and intra-frame flicker is avoided in CSK by ensuring constant average power across multiple light sources along with scrambling and high optical clock rates with megahertz.

Inter-Frame Flicker Mitigation

Inter-frame flicker mitigation applies to both data transmission (receive mode) and idle periods. While idling, idle patterns may be used to ensure light emission by the VLC transmitters have the same average brightness over adjacent MFTPs as during data transmission.

7.3   ITU-T G.9991

ITU-T G.9991: High-speed indoor visible light communication transceiver-System architecture, physical layer and data link layer specification (aka G.VLC) is a standard developed by ITU-T for indoor LOS optical networking. The high-speed indoor VLC, also known as Li-Fi, Li-Fi and Wi-Fi have different strengths, and Li-Fi's strengths provide a valuable and strong complement where Wi-Fi faces challenges. G.VLC will facilitate the acceleration and development of Li-Fi technology.

G.VLC is a recommendation proposal. It specifies the system architecture and functionality for all components of PHY and DLL (data link layer) of VLC transceivers for in-premises applications designed for high-speed optical wireless transmission (OWT) of data using LV and/or IR medium. Specifically, G.VLC defines specifications for PHY, DLL, and management layer of high-speed OWT as well as high-speed VLC system architecture and reference models.

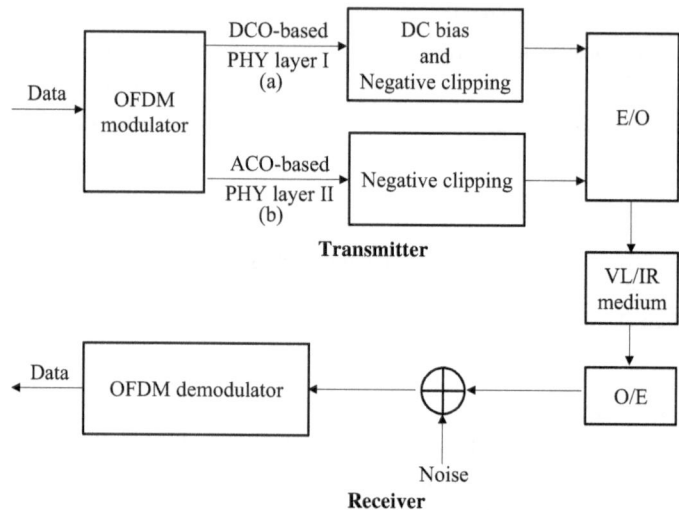

Fig. 7.12 PHY layers in G. VLC based on: **a** DCO-OFDM, and **b** ACO-OFDM

G.VLC includes two different PHY layers, as shown in Fig. 7.12: DCO-OFDM-based physical layer, it can offer very high throughput of theoretically up to 1.7 Gbps, and ACO-OFDM-based physical layer, it can cope with extreme flicker situations (e.g., deep dimming) with a penalty on achievable throughput.

In addition, DCO-OFDM is less efficient in terms of optical power than ACO-OFDM for lower SNR value. But for higher SNR values it is power efficient. This is because the DC bias used in DCO-OFDM is inefficient in terms of optical power, while the use of only half of the subcarriers to carry data in ACO-OFDM is inefficient in terms of bandwidth [9].

References

1. JEITA: *CP-1221: Visible Light Communications System* (JEITA, Tokyo 2007)
2. JEITA: *CP-1222: Visible Light ID System* (JEITA, Tokyo 2007)
3. JEITA: *CP-1223: Visible Light Beacon System* (JEITA, Tokyo 2013)
4. IEEE *Standard 802.15.7–2011: Short-Range Wireless Optical Communication Using Visible Light* (IEEE, Piscataway 2011)
5. ITU-T *G.9991: High-speed indoor visible light communication transceiver -System architecture, physical layer and data link layer specification* (ITU, Geneva 2019)
6. ITU-T *G.9991 Amendment 2: High-speed indoor visible light communication transceiver-System architecture, physical layer and data link layer specification* (ITU, Geneva 2021)
7. S. Rajagopal, R. D. Roberts, S. K. Lim: IEEE 802.15.7 visible light communication: modulation schemes and dimming support, IEEE Comm. Mag. **March 2012**, 72–82 (2012)

8. S. Berman, D. Greenhouse, I. Bailey, R. Clear, T. Raasch: Human electroretinogram responses to video displays, fluorescent lighting and other high frequency sources, Optometry and Vision Science, **68**, 645–662 (1991)
9. S. C Saju, A. J. George: Comparison of ACO-OFDM and DCO-OFDM in IM/DD systems, IJERT **4**(04), 1315–1318 (2015)

VLC Current Applications

8

The applications of VLC can be divided into two categories: terrestrial and non-terrestrial. As terrestrial applications, this chapter introduces LED-based indoor illumination-light communication systems, which including emergency-light-based LBS system, VLC-based visually impaired guidance system, Li-Fi system using VLC and PLC hybrid access mode, and LED-based IoT systems. As non-terrestrial applications, this chapter focuses on LED-based underwater optical wireless communication system with water-environment adaptive control functions, including VLC-based underwater sensor networks. As a special application, this chapter also presents examples of combining VLC with art using the color characteristics of visible light.

At present, the applications of VLC are classified mainly into two categories [1]: terrestrial and non-terrestrial. The mainstream for terrestrial application is LED-based illumination-light communication (ILC) [2, 3], including indoor information services, such as short-range one-to-one systems, Li-Fi systems [4], IoT networks [5, 6], etc., and outdoor data transmission [7], such as among signboards, streetlights, vehicles, traffic signals, and so on. The non-terrestrial applications including VL-based underwater [8] and space wireless communication systems [9, 10]. In addition, by using the color characteristics of the VL spectrum, VLC can be integrated with art to provide users with an information service environment where communication and art are shared.

In this chapter, as the VLC typical usage models of current applications, the indoor ILC system, the VLC-based blind-guidance system, the Li-Fi system, the LED-based IoT system, the underwater optical wireless communication (UOWC) system, and VLC system with art will be discussed.

8.1   Indoor One-to-One ILC System

Figure 8.1 illustrates a general scheme of an indoor VLC one-to-one link. The transmitter is an LED-lamp-based fixed device. The receiver is a desktop terminal or a portable device with a photodetector, such as a tablet terminator, a mobile phone and so on. The LED lamp (i.e., the transmitter) establishes an illumination-light link to the portable/desktop receiver. The contents for data transmission can be taken by accessing the resources on an Ethernet or storage in a built-in data memory. Typical applications include location-based services (LBSs), ID recognition systems, and audio information guides.

Figure 8.2a and b illustrate two examples of LED-based one-to-one link for an acousto-optic wireless communication system and an LBS system, respectively.

In Fig. 8.2a, the transmitter is an LED lamp using the yellow phosphor-based white LED. A light modulation and control circuit is used to transmit audio information. It is a baseband IM/DD-based system with I−4PPM scheme. The sound contents are stored in a micro memory card. The receiver is a compact batteryless acousto-optic device consisting of a solar cell, which is used for detecting signal and providing power to the receiver, and an earphone (or a speaker) for sound output [11].

Figure 8.2b illustrates an LBS/ID system based on JEITA CP-1221 and 1222 which are the VLC standards for VL ID system (Sects. 7.1.1 and 7.1.2). The transmitter is an LED lamp using the Multicolor white LED. The SC−4PPM modulation scheme is used to maintain constant transmission power and mitigate optical flicker. The frame structure for data transmission and verification follows the standard in JEITA CP−1222. The location ID information are stored in a built-in micro memory within the LED lamp. The receiver is a portable terminal (such as a tablet terminator or a mobile phone) connect with a dongle with a PD. The required location information are displayed on the tablet [12].

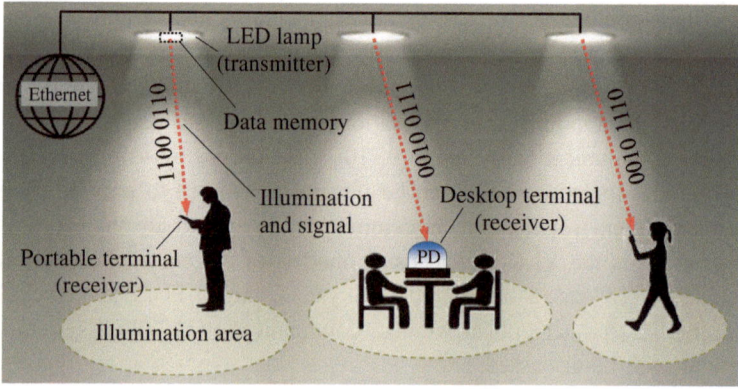

Fig. 8.1 Indoor one-to-one VLC usage model

Fig. 8.2 Examples of LED-based one-to-one link for: **a** acousto-optic wireless communication system, and **b** LBS/ID system

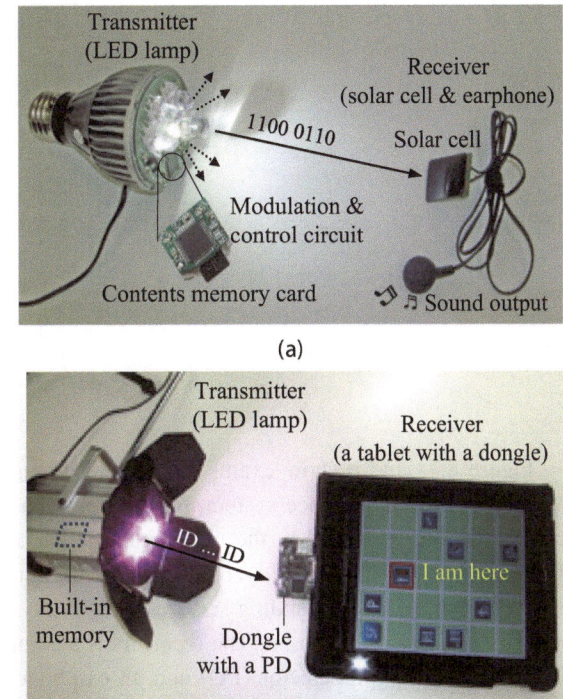

(a)

(b)

Figure 8.3 is an example of an emergency-light-based LBS system. The LED-based emergency light as a VL transmitter to send out information such as location and map to guide users to the correct location. Using the current position, location and other information to modulate the emitted light from the emergency light, users can receive these information by their portable terminal. Since emergency lights are usually always on, users can always get the required information through emergency lights.

8.2 VLC-Based Visually Impaired Guidance System

Many visually impaired individuals, said to be over 80%, are able to perceive the direction of light. X. Liu, et al., focused on this point and proposed the idea that visually impaired people who can sense light can use it by pointing an acousto-optic terminal that can be converted into a voice guidance system in the direction of the light [13]. They have set up a VLC system for visually impaired guidance by using this method. In this method, inverter fluorescent lamps were used for illumination and required data transmission, and light was received by attaching an adapter to a portable device, utilizing the speaker of the portable device.

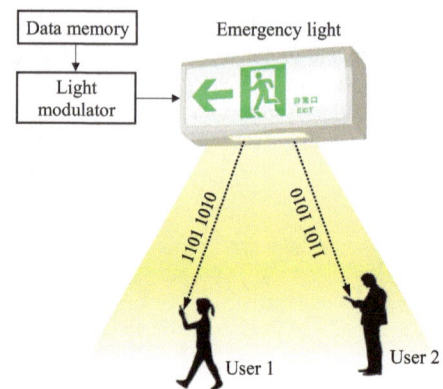

Fig. 8.3 An example of emergency-light-based LBS system

Figure 8.4a and b show examples of applications for indoor and outdoor LED-based visually impaired guidance system, respectively that is proposed by author's group.

Both in Fig. 8.4a and b, the transmitters is an LED lamp, and the Illuminating light from the LED lamp is employed for orientation guidance and information transmission, such can provide corresponding services for visually impaired, such as location guidance, obstacle reminders and so on. The receivers is a compact wearable acousto-optic terminal consisting of a photodetector (PD) and an earphone, the terminal can be attached to an assistive device for visually impaired individuals, such as a walking stick or a guide dog, the voice information received is transmitted to the user through headphones. In addition, in outdoor guidance area, multi LED lamps can be set to ensure good services.

8.3   LED-Lamp-Based Li-Fi System

A one-to-many VLC system can achieve the functionality of an illumination-light-based wireless local area network (Li-WLAN). As a complementary technique to radio-based wireless LAN Wi-Fi (wireless-fidelity), the Li-WLAN is also called Li-Fi (light-fidelity). Attractive features of the Li-Fi system are saving of frequency resources, ensuring security between adjacent WLANs.

As shown in Fig. 8.5, Li-Fi is similar to Wi-Fi in principle. Both use a wireless access point (WAP) device to connect multiple clients (user terminals) with network card functionality to form a wireless network, which is then connected to the wired Internet to achieve interoperability. The difference is that the WAP of the Li-Fi system is also a lighting fixture, known as an optical WAP (OWAP). and the information is transmitted using light (VL or/and Ir) carriers. The network card of the Li-Fi system is a bidirectional optical wireless transceiver (OWTC).

Figure 8.6 is an LED-lamp-based indoor Li-Fi system, which using VLC and PLC (power line communication) hybrid access mode to connect the Ethernet, and its data rate

Fig. 8.4 Visually impaired guidance system by using LED lighting for: **a** indoor and **b** outdoor

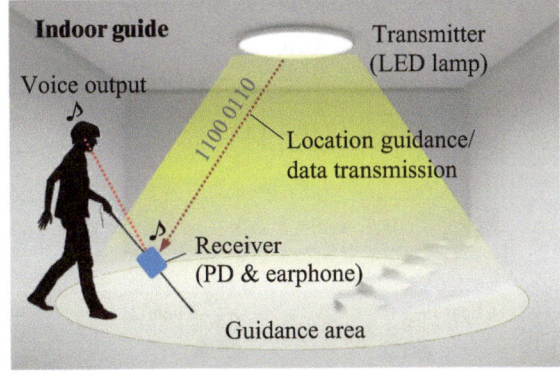

(a)

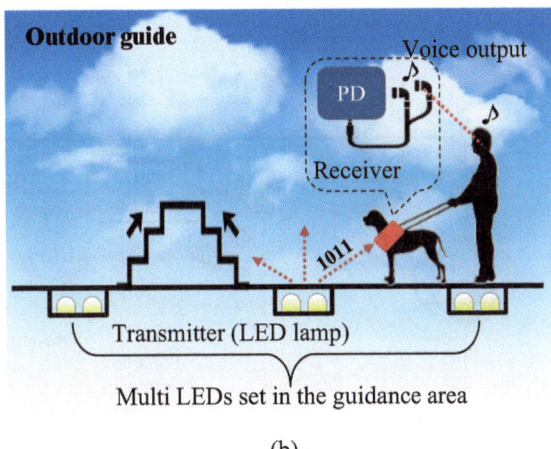

(b)

is 100Mbps. It is a multiple-access optical network, which is proposed by author' group in 2009 [14], its physical layer includes the following main aspects:

– Up/downlink carriers: since the downlink carrier must also serves as illumination, so it is VL. Using the SC-4PPM modulation scheme for flicker mitigation. The VPPM scheme, which is specified by IEEE 802. 15. 7. Can be also used when it is required both for flicker mitigation and dimming. The uplink carrier is only used for data transmission, so theoretically any carrier can be used. To avoid signal interference during bidirectional communication, different bands with wavelengths far from the downlink light center wavelength are usually used. In this example, the infrared of 860 nm is used as the uplink carrier, modulated using the 4PPM scheme specified by the IrDA (Infrared Data Association) standard.

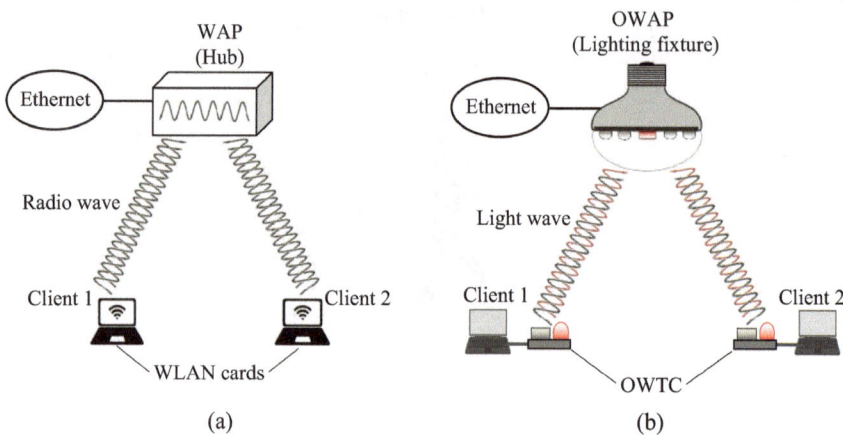

Fig. 8.5 **a** Wi-Fi and **b** Li-Fi

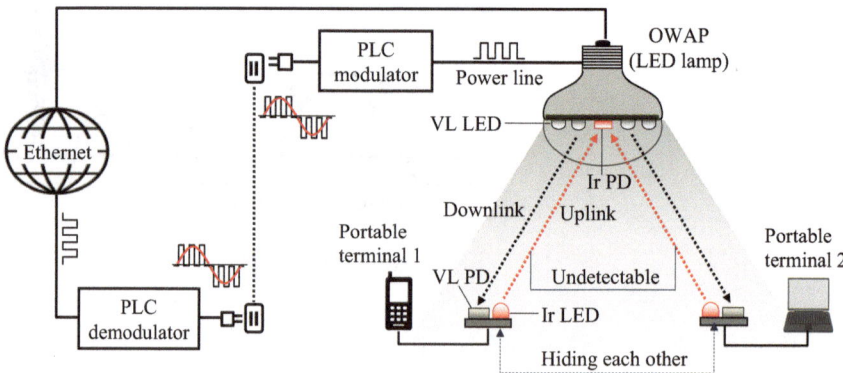

Fig. 8.6 LED-lamp-based Li-Fi system by using VLC and PLC hybrid access mode

- OWAP: using the LED lamp as the OWAP that can establish multiple links to available portable terminals, and to bridge between VL links and Ethernet LAN or achieve access to resources on an Ethernet, it is also a bidirectional transceiver, consisting of VL LEDs for illumination and sending downlink data, and an Ir PD for receiving uplink data.
- Terminal: it is a bidirectional transceiver, consisting of Ir LEDs for transmitting uplink data and a VL PD for receiving downlink data. The terminal device can be dedicated or mounted onto different portable devices, such as mobile phone, tablet, and so on.
- Data access: Direct connection to the Ethernet or via PLC modems to access the Ethernet.

And its MAC (media access control) layer includes two main aspects:

Fig. 8.7 RTS/CTS/DATA/
ACK protocol for hidden nodes

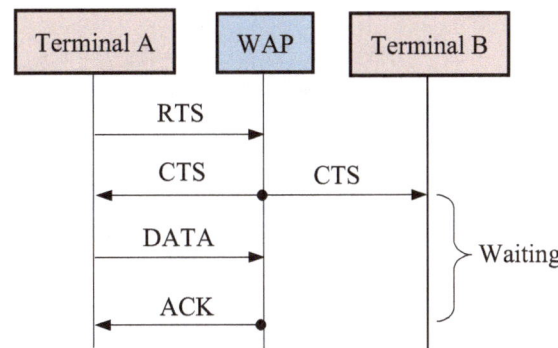

- Collision: for the *collision* problem that occurs when multiple clients (terminals) simultaneously access one host WAP, the IEEE 802.11 standard protocols for carrier-sense multiple access with collision detection (CSMA/CD) or with collision avoidance (CSMA/CA) can be referenced [15].
- Hidden nodes: for light-carrier-based Li-Fi system, *hidden nodes* may be present, i.e., terminals that are undetectable to other terminals because of the inherent directivity of light propagation, as shown in Fig. 8.6. A request to send (RTS)/clear to send (CTS) / data/acknowledge (DATA/ACK) protocol [15], which as shown in Fig. 8.7 can be used to solve the problem of the hidden nodes.

Ubiquitous is a keyword in the current information age, and the OWAP used in the Li-Fi system is a ubiquitous LED lamp, hence Li-Fi system is possible to make a common illumination environment become a ubiquitous information-service environment.

8.4   LED-Based IoT System

The VLC links also can be employed to achieve direct, peer-to-peer communication between plural portables and/or plural fixed terminals, this type of ad hoc interconnection is well suited to a ubiquitous Internet of things (IoT).

The IoT refers to a network of interconnected various physical devices, which it extends the terminals relative to *people* in WLAN to *things*, and then uses these things with independent functions in the real environment as sub-nodes, connecting them together with a main node, such as a WAP to achieve interconnection, information exchange, and sharing. The communication between the main node and the sub nodes is bidirectional wireless links.

LED-based IoT system uses LED illumination light as the carrier, with lighting fixtures equipped with OWAP functionality as the main node of the IoT. The core and foundation of LED-based IoT still rely on illumination-light WLAN, i.e., the Li-Fi network.

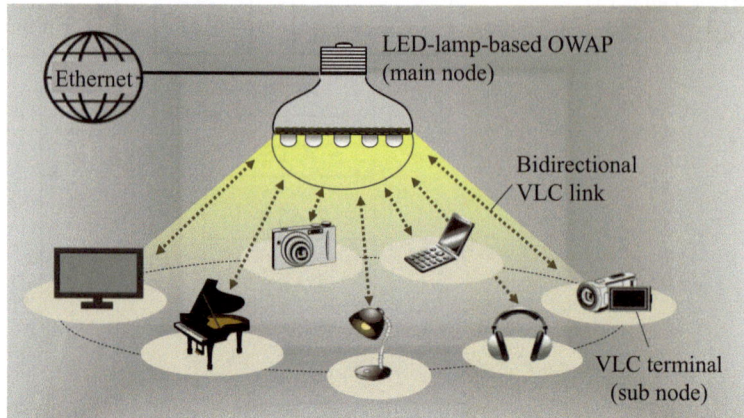

Fig. 8.8 VLC-based indoor IoT

Figures 8.8 and 8.9 shown indoor and outdoor LED-based IoT systems, respectively, they merge lighting and data communications in applications such as LED lamps, home appliances for indoor, and signboards, streetlights, vehicles, traffic signals, and so on for outdoor.

The technical contents of realizing VLC-based IoT include:

– Physically, enabling each object involved in the IoT to have bidirectional optical wireless communication and self-localization/self-configuration capabilities.
– Logically, assigning a unique identifier or address to each object in the IoT.
– Visualization of data/information to provide convenience for human participation and control.
– Development of communication protocols. There are many different types of *things* that IoT terminals connect to, and there is currently no unified specification that adapts to multiple *things*, which is also one of the challenges faced by optical IoT systems. In addition, the networking of the light sources as well as upgrading current infrastructures to support VLC-based IoT is another challenge, which requires support from the lighting industry.

Despite its numerous benefits, IoT also poses challenges, such as privacy concerns, interoperability issues, and security vulnerabilities. As technology continues to advance, IoT is expected to play an increasingly significant role in shaping the way we live, work, and interact with the world around us.

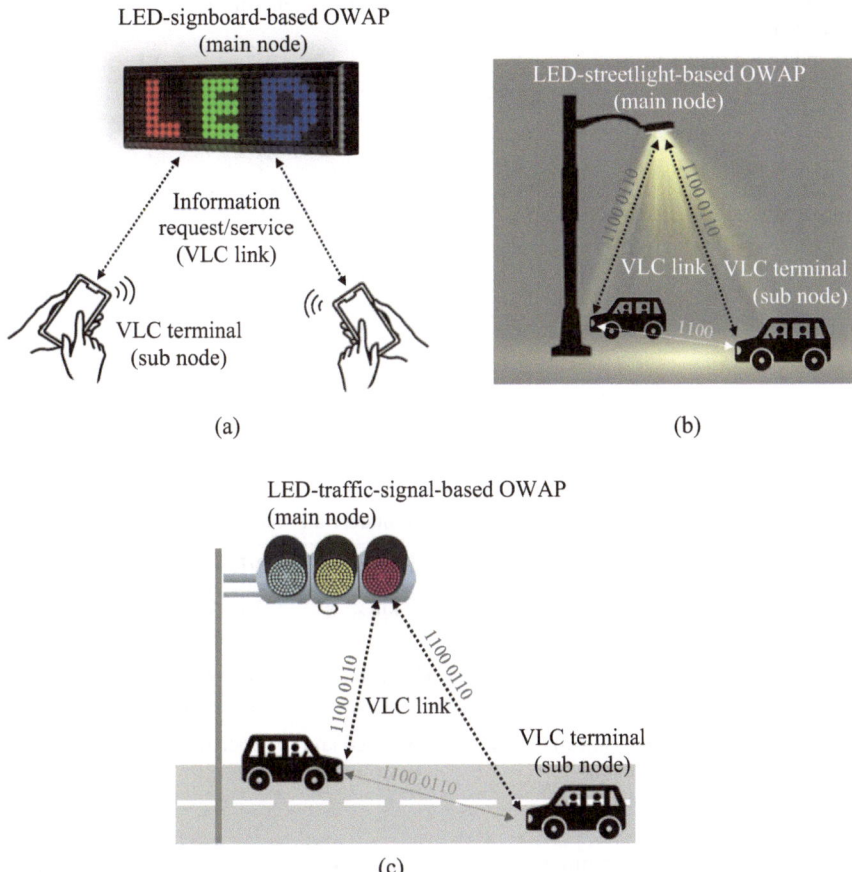

Fig. 8.9 VLC-based outdoor IoT by using LED-based: **a** signboard, **b** streetlight, and **c** traffic signals

8.5 LED-Based UOWC System

The LED-based UOWC is a typical application of VLC-based non-terrestrial system. Comparisons of the absorption coefficients and carrier characteristics of underwater radio, sound, and VL waves are shown in Fig. 8.10 and given in Table 8.1 [16, 17].

The sound-wave communication technique is used in almost all present commercial underwater data transmission systems as it can propagate well in seawater and can reach far distances (up to several kilometers). Unfortunately, sound-wave communication systems do not enable large-capacity links, such those needed for image data transmissions, because they are inherently low speed and a Doppler effect is caused by this low speed.

Fig. 8.10 Absorption coefficients of electromagnetic waves and sound waves in water

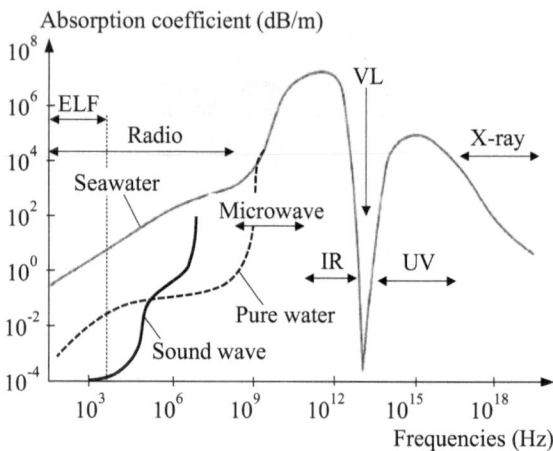

Table 8.1 Comparisons of carrier characteristics of underwater radio, sound, and VL waves

	Radio wave	Sound wave	VL wave
Distance	<1 m in ELF	>10 km	10–100 m (depending on the seawater turbidity)
Speed	Extremely low speed	Low speed (<1500 m/s)	High speed (large capacity)
Problem	Very low speed so cannot use terrestrial technique	Low speed and doppler effect	Spectral attenuation

Moreover, when the communication system is dynamic, severe frequency-dependent dispersion may arise even at short ranges.

An alternative to sound-wave communication is using radio waves at extremely low frequencies (ELF: <10 kHz) due to their low absorption coefficient in pure seawater, as shown in Fig. 8.10. Despite this, radio-wave-based underwater communication still presents some intrinsic drawbacks. In fact, the major obstacle in using radio for underwater communication is the severe attenuation due to the conducting nature of seawater. In particular, the attenuation is very high for high-frequency radio waves and, since the current terrestrial technology for wireless communication is often based on high frequencies in the order of Gbps, it is practically impossible to use terrestrial techniques in underwater applications.

VL-based underwater communication techniques are being considered as a possible solution to this because of their large data capacity due to the inherent wide bandwidth. Moreover, seawater exhibits a window of reduced absorption in the visible-spectrum range; particularly light wavelengths between 400−650 nm, where water is relatively transparent to light and absorption takes its minimum value, as shown in

Fig. 8.10. Nevertheless, for different marine environments and turbidities, VL will undergo wavelength-dependent attenuation in natural water, as shown in Fig. 8.11. For pure seawater type of Fig. 8.11a, the absorption is dominated almost by the attenuation of seawater molecule, the region of ideal wavelength with lowest attenuation for visible-light propagation is within the blue-green band between 400 and 500 nm. In case of high-turbidity bay water of Fig. 8.11b, the total absorption in visible-light band is dominated by the combination of organic and inorganic particles, the ideal transmission wavelength is shifted from blue-green towards green-yellow band around 550−600 nm.

Figure 8.12 illustrates the behaviours of particles in seawater when visible-light wave propagates in seawater. The particles are not easily hit by red light with long wavelength, Fig. 8.12a, but easily by blue light with short wavelength, as shown in Fig. 8.12b. When light wave hits a particle, its intensity will be attenuation because scatter of light. That is why the red light with longer wavelength has better transmittance then blue light with shorter wavelength in high-turbidity seawater such as the bay.

Adaptive control techniques can be applied to help overcome the seawater turbidity of the spatiotemporal change to obtain efficient and reliable light propagations over larger

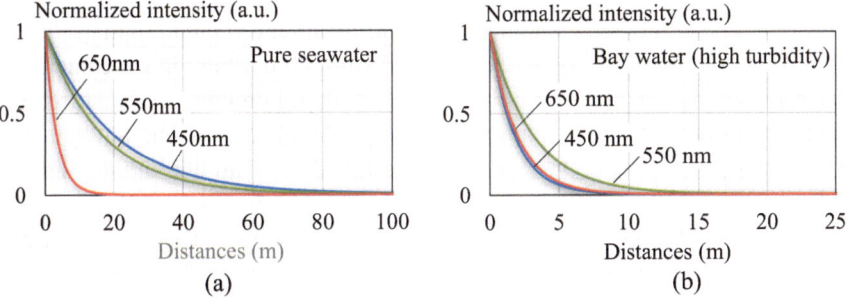

Fig. 8.11 Spectrum-intensity attenuation for different seawater types in horizontal direction: **a** pure seawater and **b** bay water

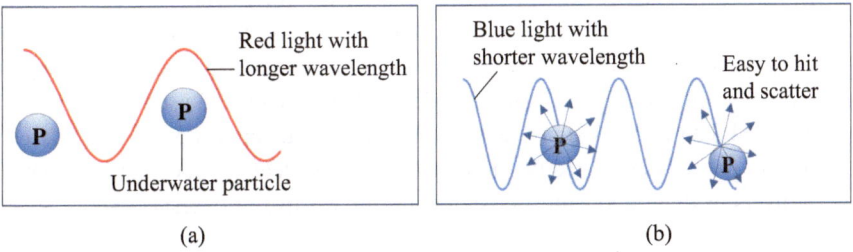

Fig. 8.12 Influences of underwater particles on VL transmittance: **a** red light and **b** blue light

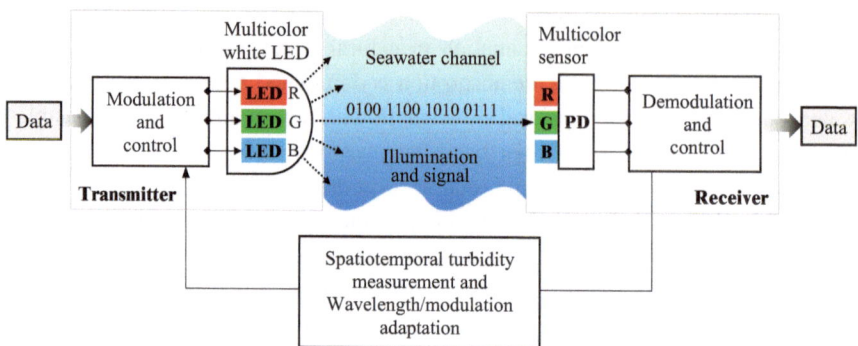

Fig. 8.13 LED-based adaptive UOWC system

ranges and construct a robust link. Figure 8.13 shows an LED-based adaptive UOWC system with wavelength- and modulation-adaptation control.

The link between the transmitter and the receiver is IM/DD type. The LPPM scheme is used to perform modulation-adaptation control. As Sect. 5.2.1 mentioned, for LPPM, the SNR requirement to reach a certain BER is decreased with increasing L, hence in cases of longer-distance turbid seawater channels, by adjusting L communication quality can be ensured. However, note that an increase in the L causes a decrease in the communication speed. Alternatively, in order to reach wavelength-adaptation control, a multicolor white LED can be used. Each color chip has an independent and different wavelength peak, which can act as a separate channel and is controlled for seawater turbidity adaptation. At the receiving side, a multicolor sensor is used to receive different-color light from each wavelength channel. Each color channel can continuously measure the intensity of monochromatic light in the color channel. The measurement results inform which color of light is selected for data transmission. This optical measurement can be achieved by automatically switching the three-color PDs.

As shown in Fig. 8.14a, the UOWC can be used to construct an underwater optical wireless sensor network (UOWSN) [18, 19]. In this case, the links between each subnode are bidirectional VL wireless channels, the distances are about from 3 to 10 m. The space division and visibility of VL can ensure each subnode is independent and identifiable both in space and time. These subnode data are transmitted long-distance to a terrestrial station by using optical fibers via a main node. UOWC also can be used for short-distance and high-speed data transmissions between an autonomous underwater vehicle (AUV) or a remotely operated vehicle (ROV) and an underwater detector [20], and conversations between divers or diver and sailor, as shown in Fig. 8.14b and c, respectively.

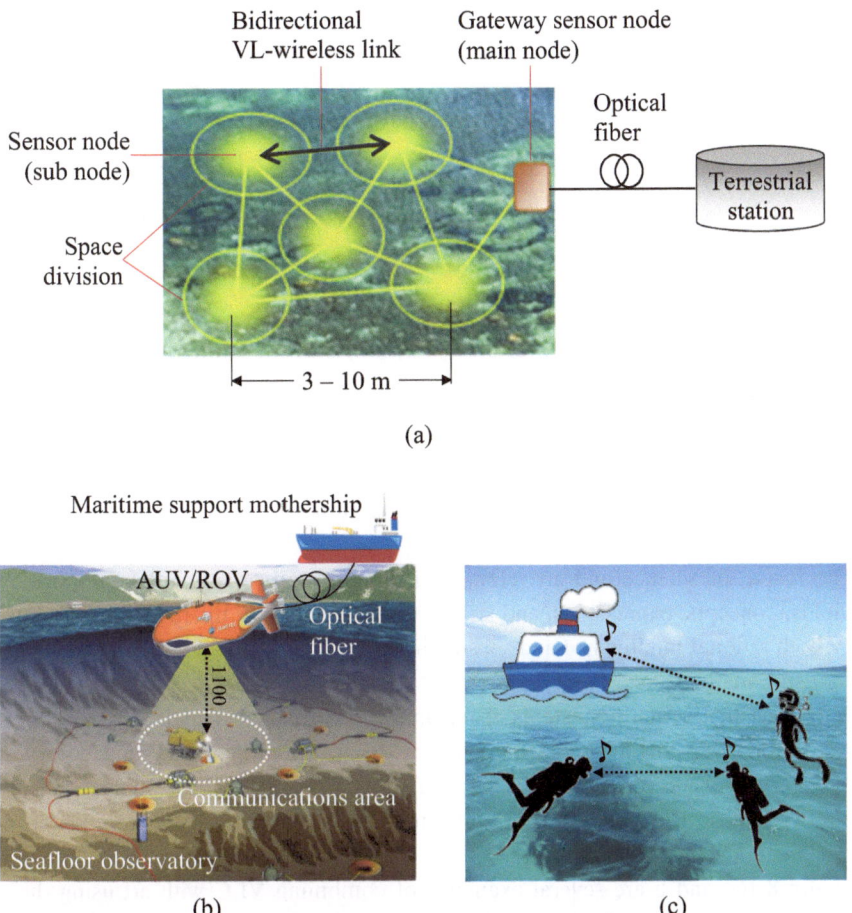

Fig. 8.14 Applications of UOWC for: **a** the UOWSN, **b** underwater data transmissions, and **c** underwater conversations

8.6   Fusion of VLC and Art

Science and art, at their core, are both methods of exploring the essence of the world. While science seeks to understand and explain phenomena through rigorous observation, experimentation, and analysis, art explores the world through subjective interpretation, emotions, and sensory experiences. Both disciplines strive to understand the world and communicate their discoveries to others, but by different means:

Science: inducing natural phenomena into mathematical models, studying the mechanisms of models, manufacturing generalized machinery, emphasizing the intellect of machinery, and creating economic civilization;

Fig. 8.15 Example of fusing art with mathematics

$$x^2 + \left(y - x^{\frac{2}{3}}\right)^2 = 1$$

Art: abstracting human thoughts into sensory models, expressing and transcending models, creating generalized beauty, emphasizing human sensation, and creating spiritual civilization.

The fusion of science and art refers to the integration or/and collaboration between the fields, concepts, expressions, and so on of science and art, the result makes people to enjoy both rational intelligence and sensual beauty. For example, Fig. 8.15 is a generative art by using mathematics.

People aesthetic intuition stems from appropriately shaped and colored effects. In VLCs, color is an inherent characteristic of VL in scientific sense, and color is also the source of visual art. The color characteristic of VL makes VLC more conducive to integration with art, providing users with an information service environment for data communications and artistic sharing.

Figure 8.16a and b are several examples of combining VLC with art using the color characteristics of VL. In Fig. 8.16a, LED decorative lamps with *bud*, and *egg* (i.e., a symbol of life) shapes respectively emit colored light corresponding to the colors of buds and eggs, and then modulate and upload data (voice, image, etc.) about *buds* or/and *life* onto the corresponding colors of light for transmission, giving users an immersive experience. In the example of Fig. 8.16b, utilizing the colors of VL and employing MWD techniques, a colorful multi-information service environment has been constructed. In this environment, users can enjoy different information services by selecting lights of different colors. For example, blue light corresponds to a piece of *music*, green light corresponds to a scenic *view*, and red light corresponds to a *story*, and so on. Users will enjoy the pleasure of scientific selection/transformation while satisfying the artistic beauty of color.

Overall, the fusion of science and art can foster interdisciplinary collaboration, enhance public engagement with science, inspire creativity and innovation, and promote a deeper understanding of the natural world and human experience.

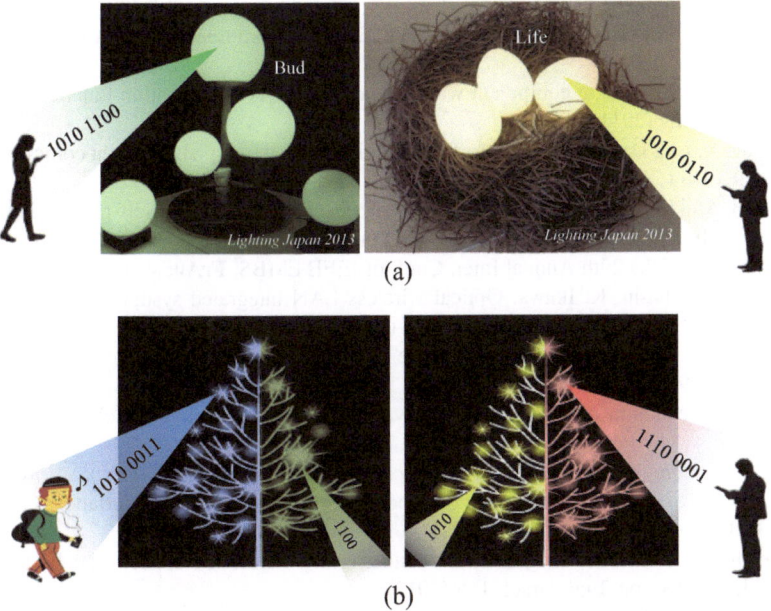

Fig. 8.16 Examples of combining VLC with art

References

1. X. Lin: Visible-light wireless communications technique using LED lighting, IEICE Tech. Rep. **115**(247), 63–68 (2015)
2. X. Lin: Optical wireless ubiquitous information service using LED lighting, Mon. Disp. **18**(10), 46–52 (2012)
3. X. Lin: Chapter 9.2, LED-Based Illumination-Light Communication Device (Technical Information Institute, Tokyo 2014), pp. 609–614
4. R. Hassan, M. S. Flayyih, A. Mahdi, A. Inn, A. S. Sadeq, D. F. Murad: Visible light communication technology for data transmission using Li-Fi, 2020 2nd ICCIS, https://doi.org/10.1109/ICCIS49240.2020.9257654
5. M. Oshima, H. Aoyama, K. Nakanishi, T. Maeda: Image-sensor based visible light communication for IoT, OPTRONICS, No. 8, 64–71 (2017)
6. N. Iizuka: Industrial IoT with image sensor communication, OPTRONICS, No. 8, 78–83 (2017)
7. K. Yutaka, T. Nakama: Outdoor LED communication technology using light space modulation element, Jap. Jour. of Opt., **45**(2), 68–70 (2014)
8. X. Lin: Underwater wireless communication system of adaptation wavelength using visible light, IEICE Trans. Fundam. E100-A(1), 185–193 (2017)
9. M. Toyoshima: R&D trends toward practical realization of space laser communications, Jap. Jour. of Opt., **45**(2), 62–67 (2014)

10. T. Araki: A brief of space optical communication and expectation to visible-light communication technology for space system, Proc. of IEICE general conference 2016, **AS-3–4**, S-41–S42 (2016)
11. X. Lin, H. Itoh: LED light equipment with optical wireless communication functions. In: Proc. World Eng. Conv (2011)
12. Y. Yasuda, X. Lin: Development of the facility guidance system by visible light communication, Proc. IEIEJ I–11, 467–470 (2013)
13. X. Liu, H. Makino, S. Kobayashi and Y. Maeda: Design of an indoor self-positioning system for the visually impaired -simulation with RFID and Bluetooth in a visible light communication system-, Proc. of the 29th Annual Inter. Conf. of IEEE EMBS, **FrA05.4**, 1655–1658 (2007)
14. X. Lin, K. Hirohashi, K. Ikawa: Optical wireless LAN integrated systems with illumination function, IEICE Tech. Rep. **109**(400), 43–48 (2010)
15. IEEE: Standard 802.11: Wireless LAN (The Institute of Electrical and Electronic Engineers, Piscataway 1998)
16. S. Nakao: Attenuation of electromagnetic waves in seawater, Def. Technol. 1987(9), 22–30 (1987)
17. X. Lin, T. Matsumura: Visible light communications, B. Mukherjee et al. (Eds.), *Springer Handbook of Optical Networks*, Springer Handbooks, Springer Nature Switzerland AG, **Part D** (35.5.3), 1121–1122 (2020)
18. L. Ghelardoni, A. Ghio, D. Anguita: Smart underwater wireless sensor networks. In: IEEE 27th Conv. Electr. Electron. Eng. Israel, 1–5 (2012)
19. X. Lin: Wavelength-adaptation underwater optical wireless sensor network using visible light communication, Adv. in Comp. and Comm., **1**(1), 7–15 (2020) http://dx.doi.org/10.26855/acc.2020.12.002
20. T. Sawa, X. Lin: Research of underwater optical wireless robust communication, Jap. Jour. of Opt., 45(2), 55–61 (2014)